T0258893

Thermal-Aware Testing of Digital VLSI Circuits and Systems

Thermal-Aware Testing of Digital VLSI Circuits and Systems

Santanu Chattopadhyay

CRC Press
Taylor & Francis Group
Boca Raton London New York

CRC Press is an imprint of the
Taylor & Francis Group, an **Informa** business

CRC Press
Taylor & Francis Group
6000 Broken Sound Parkway NW, Suite 300
Boca Raton, FL 33487-2742

© 2018 by Taylor & Francis Group, LLC
CRC Press is an imprint of Taylor & Francis Group, an Informa business

No claim to original U.S. Government works

Printed on acid-free paper

International Standard Book Number-13: 978-0-8153-7882-2 (Hardback)

Library of Congress Cataloging-in-Publication Data

Names: Chattopadhyay, Santanu, author.
Title: Thermal-aware testing of digital VLSI circuits and systems / Santanu Chattopadhyay.
Description: First edition. | Boca Raton, FL : Taylor & Francis Group, CRC Press, 2018. | Includes bibliographical references and index.
Identifiers: LCCN 2018002053| ISBN 9780815378822 (hardback : acid-free paper) | ISBN 9781351227780 (ebook)
Subjects: LCSH: Integrated circuits--Very large scale integration--Testing. | Digital integrated circuits--Testing. | Integrated circuits--Very large scale integration--Thermal properties. | Temperature measurements.
Classification: LCC TK7874.75 .C464 2018 | DDC 621.39/50287--dc23
LC record available at https://lccn.loc.gov/2018002053

Visit the Taylor & Francis Web site at
http://www.taylorandfrancis.com

and the CRC Press Web site at
http://www.crcpress.com

To
SANTANA, MY WIFE
My Inspiration
and
SAYANTAN, OUR SON
Our Hope

———————————

Contents

List of Abbreviations

3D	Three dimensional
ATE	Automatic Test Equipment
ATPG	Automatic Test Pattern Generation
BIST	Built-In Self Test
CAD	Computer-Aided Design
CLK	Clock Signal
CTM	Compact Thermal Model
CUT	Circuit Under Test
DFT	Design for Testability
DI	Data Input
DPSO	Discrete Particle Swarm Optimization
FEM	Finite Element Method
IC	Integrated Circuit
IP	Intellectual Property
LFSR	Linear Feedback Shift Register
LT-RTPG	Low Transition-Random Test Pattern Generator
MISR	Multiple-Input Signature Register
MTTF	Mean Time to Failure
NoC	Network-on-Chip
NTC	Node Transit Count
ORA	Output-Response Analyzer
PCB	Printed Circuit Board
PSO	Particle Swarm Optimization
SE	Scan Enable
SI	Scan Input

SNM	Static Noise Margin
SoC	System-on-Chip
SSI	Small-Scale Integration
TAM	Test Access Mechanism
TAT	Test Application Time
TI	Thermal Interface
TPG	Test Pattern Generator
VLSI	Very Large-Scale Integration
WTC	Weighted Transition Count

Preface

Demand for improved system performance from silicon integrated circuits (ICs) has caused a significant increase in device density. This, associated with the incorporation of power-hungry modules into the system, has resulted in power consumption of ICs going up by leaps and bounds. Apart from threatening to violate the power-limits set by the designer, the process poses a formidable challenge to the test engineer as well. Due to the substantially higher switching activity of a circuit under test (CUT), average test power is often 2X higher than the normal mode of operation, whereas the peak power can be up to 30X higher. This excessive test power consumption not only increases the overall chip temperature, but also creates localized overheating hot spots. The test power minimization techniques do not necessarily lead to temperature minimization. Temperature increase is a local phenomenon and depends upon the power consumption, as well as heat generation of surrounding blocks. With an increase in temperature, leakage current increases, causing a further increase in power consumption and temperature. Thus, thermal-aware testing forms a discipline by itself. The problem can be addressed both at the circuit level and at the system level. While the circuit-level techniques address the issues of reducing the temperature of individual circuit modules within a chip, system-level ones deal with test scheduling problems. Typical circuit-level techniques include test-vector reordering, don't care bit filling, scan chain structuring, etc. System-level tools deal with scheduling of core tests and test-data compression in System-on-Chip (SoC) and Network-on-Chip (NoC) designs. This book highlights the research activities in the domain of thermal-aware testing. Thus, this book is suitable for researchers working on power- and thermal-aware design and testing of digital very large scale integration (VLSI) chips.

Organization: The book has been organized into five chapters. A summary of the chapters is presented below.

Chapter 1, titled "VLSI Testing—An Introduction," introduces the topic of VLSI testing. The discussion includes importance of testing in the VLSI design cycle, fault models, test-generation techniques, and design-for-testability (DFT) strategies. This has been followed by the sources of power dissipation during testing and its effects on the chip being tested. The problem of thermal-aware testing has been enumerated, clearly bringing out the limitations of power-constrained test strategies in reducing peak temperature and its variance. The thermal model used in estimating temperature values has been elaborated.

Chapter 2, "Circuit Level Testing," notes various circuit-level techniques to reduce temperature. Reordering the test vectors has been shown to be a potential avenue to reduce temperature. As test-pattern generation tools leave large numbers of bits as *don't cares*, they can be filled up conveniently to aid in temperature reduction. Usage of different types of flip-flops in the scan chains can limit the activities in different portions of the circuit, thus reducing the heat generation. Built-in self-test (BIST) strategies use an on-chip test-pattern generator (TPG) and response analyzer. These modules can be tuned to get a better temperature profile. Associated techniques, along with experimental results, are presented in this chapter.

Chapter 3 is titled as "Test Data Compression." To reduce the transfer time of a test pattern, test data are often stored in the tester in a compressed format, which is decompressed at the chip level, before application. As both data compression and temperature minimization strategies effectively exploit the *don't care* bits of test patterns, there exists a trade-off between the degree of compression and the attained reduction in temperature. This chapter presents techniques for dictionary-based compression, temperature-compression trade-off, and temperature reduction techniques without sacrificing on the compression.

Chapter 4, titled "System-on-Chip Testing," discusses system-level temperature minimization that can be attained via scheduling the tests of various constituent modules in a system-on-chip (SoC). The principle of superposition is utilized to get the combined effect of heating from different sources onto a particular module. Test scheduling algorithms have been reported based on the superposition principle.

Chapter 5, titled as "Network-on-Chip Testing," discusses thermal-aware testing problems for a special variant of system-on-chip (SoC), called network-on-chip (NoC). NoC contains within it a message transport framework between the modules. The framework can also be used to transport test data. Optimization algorithms have been reported for the thermal-aware test scheduling problem for NoC.

Santanu Chattopadhyay
Indian Institute of Technology
Kharagpur

Acknowledgments

I MUST ACKNOWLEDGE THE CONTRIBUTION of my teachers who taught me subjects such as Digital Logic, VLSI Design, VLSI Testing, and so forth. Clear discussions in those classes helped me to consolidate my knowledge in these domains and combine them properly in carrying out further research works in digital VLSI testing. I am indebted to the Department of Electronics and Information Technology, Ministry of Communications and Information Technology, Government of India, for funding me for several research projects in the domain of power- and thermal-aware testing. The works reported in this book are the outcome of these research projects. I am thankful to the members of the review committees of those projects whose critical inputs have led to the success in this research work. I also acknowledge the contribution of my project scholars, Rajit, Kanchan, and many others in the process.

My source of inspiration for writing this book is my wife Santana, whose relentless wish and pressure has forced me to bring the book to its current shape. Over this long period, she has sacrificed a lot on the family front to allow me to have time to continue writing, taking all other responsibilities onto herself. My son, Sayantan always encouraged me to write the book.

I also hereby acknowledge the contributions of the publisher, CRC Press, and its editorial and production teams for providing me the necessary support to see my thoughts in the form of a book.

Santanu Chattopadhyay

Author

Santanu Chattopadhyay received a BE degree in Computer Science and Technology from Calcutta University (BE College), Kolkata, India, in 1990. In 1992 and 1996, he received an MTech in computer and information technology and a PhD in computer science and engineering, respectively, both from the Indian Institute of Technology, Kharagpur, India. He is currently a professor in the Electronics and Electrical Communication Engineering Department, Indian Institute of Technology, Kharagpur. His research interests include low-power digital circuit design and testing, System-on-Chip testing, Network-on-Chip design and testing, and logic encryption. He has more than one hundred publications in international journals and conferences. He is a co-author of the book *Additive Cellular Automata—Theory and Applications*, published by the IEEE Computer Society Press. He has also co-authored the book titled *Network-on-Chip: The Next Generation of System-on-Chip Integration*, published by the CRC Press. He has written a number of text books, such as *Compiler Design, System Software*, and *Embedded System Design*, all published by PHI Learning, India. He is a senior member of the IEEE and also one of the regional editors (Asia region) of the IET Circuits, Devices and Systems journal.

CHAPTER 1

VLSI Testing

An Introduction

M OORE'S LAW [1] HAS been followed by the VLSI chips, doubling the complexity almost every eighteen months. This has led to the evolution from SSI (*small-scale integration*) to VLSI (*very large-scale integration*) devices. Device dimensions, referred to as *feature size*, are decreasing steadily. Dimensions of transistors and interconnects have changed from tens of microns to tens of nanometers. This reduction in feature size of devices has resulted in increased frequency of operation and device density in the silicon floor. This trend is likely to continue in the future. However, the reduction in feature size has increased the probability of manufacturing *defects* in the IC (integrated circuit) that results in a *faulty* chip. As the feature size becomes small, a very small defect may cause a transistor or an interconnect to fail, which may lead to total failure of the chip in the worst case. Even if the chip remains functional, its operating frequency may get reduced, or the range of functions may get restricted. However, defects cannot be avoided because the silicon wafer is never 100% pure, making devices located at impurity sites malfunction. In the VLSI manufacturing process, a large number of chips are produced on the same silicon wafer. This reduces the cost of production for individual chips, but each chip needs to be tested separately—checking one of the lot does not give the guarantee of correctness for the others. Testing is necessary at other stages of the manufacturing process as well. For example, an electronic system consists of *printed circuit boards* (PCBs). IC chips are mounted on PCBs and interconnected via metal lines. In the system design process, the *rule of ten* says that the cost of detecting a faulty IC increases by an order of magnitude as it progresses through each stage of the manufacturing process— device to board to system to field operation. This makes testing a very important operation to be carried out at each stage of the manufacturing process. Testing also aids in improving process yield by analyzing the cause of defects when faults are encountered. Electronic equipment, particularly that used in safety-critical applications (such as medical electronics), often requires periodic testing. This ensures fault-free operation of such systems and helps to initiate repair procedures when faults are detected. Thus, VLSI testing is essential for designers, product engineers, test engineers, managers, manufacturers, and also end users.

The rest of the chapter is organized as follows. Section 1.1 presents the position of testing in the VLSI design process. Section 1.2 introduces commonly used fault models. Section 1.3 enumerates the deterministic test-generation process. Section 1.4 discusses *design for testability* (DFT) techniques. Section 1.5 presents the sources of power dissipation during testing and associated concerns. Section 1.6 enumerates the effects of high temperature in ICs. Section 1.7 presents the thermal model. Section 1.8 summarizes the contents of this chapter.

1.1 TESTING IN THE VLSI DESIGN PROCESS

Testing essentially corresponds to the application of a set of test stimuli to the inputs of a *circuit under test* (CUT) and analyzing the responses. If the responses generated by the CUT are correct, it is said to pass the test, and the CUT is assumed to be fault-free. On the other hand, circuits that fail to produce correct responses for any of the test patterns are assumed to be faulty. Testing is carried out at different stages of the life cycle of a VLSI device.

Typically, the VLSI development process goes through the following stages in sequence: design, fabrication, packaging, and quality assurance. It starts with the specification of the system. Designers convert the specification into a VLSI design. The design is verified against the set of desired properties of the envisaged application. The verification process can catch the design errors, which are subsequently rectified by the designers by refining their design. Once verified and found to be correct, the design goes into fabrication. Simultaneously, the test engineers develop the test plan based upon the design specification and the fault model associated with the technology. As noted earlier, because of unavoidable statistical flaws in the silicon wafer and masks, it is impossible to guarantee 100% correctness in the fabrication process. Thus, the ICs fabricated on the wafer need to be tested to separate out the defective devices. This is commonly known as *wafer-level testing*. This test process needs to be very cautious as the bare-minimum die cannot sustain high power and temperature values. The chips passing the wafer-level test are packaged. Packaged ICs need to be tested again to eliminate any devices that were damaged in the packaging process. Final testing is needed to ensure the quality of the product before it goes to market; it tests for parameters such as timing specification, operating voltage, and current. Burn-in or stress testing is performed in which the chips are subjected to extreme conditions, such as high supply voltage, high operating temperature, etc. The burn-in process accelerates the effect of defects that have the potential to lead to the failure of the IC in the early stages of its operation.

The quality of a manufacturing process is identified by a quantity called *yield*, which is defined as the percentage of acceptable parts among the fabricated ones.

$$Yield = \frac{Parts\ accepted}{Parts\ fabricated} \times 100\%$$

Yield may be low because of two reasons: random defects and process variations. Random defects get reduced with improvements in *computer aided design* (CAD) tools and the VLSI fabrication process. Hence, parametric variations due to process fluctuation become the major source of yield loss.

Two undesirable situations in IC testing may occur because of the poorly designed test plan or the lack of adherence to the *design for testability* (DFT) policy. The first situation is one in which a faulty device appears to be good and passes the test, while in the second case, a good chip fails the test and appears to be faulty. The second case directly affects the yield, whereas the first one is more serious because those faulty chips are finally going to be rejected during the field deployment and operation. *Reject rate* is defined as the ratio of field-rejected parts to all parts passing the quality test.

$$Reject\ rate = \frac{(Faulty\ parts\ passing\ final\ test)}{(Total\ number\ of\ parts\ passing\ final\ test)}$$

In order to test a circuit with n inputs and m outputs, a predetermined set of input patterns is applied to it. The correct responses corresponding to the patterns in this set are precomputed via circuit simulation. These are also known as golden responses. For the *circuit under test* (CUT), if a response corresponding to any of the applied input patterns from the set does not match with the golden response, the circuit is said to be faulty. Each such input pattern is called a *test vector* for the circuit, and the whole set is called a *test-pattern set*. It is expected that, in the presence of faults, the output produced by the circuit on applying the test-pattern set will differ from the golden response for at least one pattern. Naturally, designing a good test-pattern set is a challenge. In a very simplistic approach, for an n-input CUT, the test-pattern set can contain all the 2^n possible input patterns in it. This is known as *functional testing*. For a combinational circuit, functional testing literally checks its truth table. However, for a sequential circuit, it may not ensure the testing of the circuit in all its states. Further, with the increase in the value of n, the size of the test-pattern set increases exponentially.

Another mode of testing, called *structural testing*, uses the structural information of the CUT and a *fault model*. A fault model, discussed later, abstracts the physical defects in the circuit into different types of faults. Test vectors are generated targeting different types of faults in the circuit elements—gates, transistors, flip-flops, and signal lines. A quantitative measure about the quality of a test-pattern set corresponding to a fault model is expressed as *fault coverage*. *Fault coverage* of a test-pattern set for a circuit with respect to a fault model is defined as the ratio of the number of faults detected by the test set to the total number of modeled faults in the circuit. However, for a circuit, all faults may not be detectable. For such an undetectable fault, no pattern exists that can produce two different outputs from the faulty and fault-free circuits. Determining undetectable faults for a circuit is itself a difficult task. *Effective fault coverage* is defined as the ratio of the number of detected faults to the total number of faults less the number of undetectable faults. *Defect level* is defined as,

$$Defect\ level = 1 - yield^{(1-fault\ coverage)}.$$

Improving fault coverage improves defect level. Since enhancing yield may be costly, it is desirable to have test sets with high fault coverage.

1.2 FAULT MODELS

As the types of defects in a VLSI chip can be numerous, it is necessary to abstract them in terms of some faults. Such a fault model should have the following properties:

1. Accurately reflect the behavior of the circuit in the presence of the defect.

2. Be computationally efficient to generate test patterns for the model faults and to perform fault simulation for evaluating the fault coverage.

Many fault models have been proposed in the literature; however, none of them can comprehensively cover all types of defects in VLSI chips. In the following, the most important and widely used models have been enumerated.

1.2.1 Stuck-at Fault Model

A stuck-at fault affects the signal lines in a circuit such that a line has its logic value permanently as 1 (*stuck-at-one* fault) or 0 (*stuck-at-zero* fault), irrespective of the input driving the line. For example, the output of a 2-input AND-gate may be stuck-at 1. Even if one of the inputs of the AND-gate is set to zero, the output remains at 1 only. A stuck-at fault transforms the correct value on the faulty signal line to appear to be stuck at a constant logic value, either 0 or 1. A *single stuck-at fault* model assumes that only one signal line in the circuit is faulty. On the other hand, a more generic *multiple stuck-at fault* model assumes multiple lines become faulty simultaneously. If there are n signal lines in the circuit, in a single stuck-at fault model, the probable number of faults is $2n$. For a multiple stuck-at fault model, the total number of faults becomes 3^n-1 (each line can be in one of the three states— fault free, stuck-at 1, or stuck-at 0). As a result, multiple stuck-at fault is a costly proposition as far as test generation is concerned. Also, it has been observed that test patterns generated assuming a single stuck-at fault model are often good enough to identify circuits with multiple stuck-at faults also (to be faulty).

1.2.2 Transistor Fault Model

While the stuck-at fault model is suitable for gate-level circuits, at switch level, a transistor may be stuck-open or stuck-short. Because a gate consists of several transistors, stuck-at faults at input/output lines of gates may not be sufficient to model the behavior of the gate if one or more transistors inside the gate are open or short. Detecting a stuck-open fault often requires a sequence of patterns to be applied to the gate inputs. On the other hand, stuck-short faults are generally detected by measuring the current drawn from the power supply in the steady-state condition of gate inputs. This is more commonly known as I_{DDQ} testing.

1.2.3 Bridging Fault Model

When two signal lines are shorted due to a defect in the manufacturing process, it is modeled as a bridging fault between the two. Popular bridging fault models are *wired-AND* and *wired-OR*. In the *wired-AND* model, the signal net formed by the two shorted lines take the value, logic 0, if either of the lines are at logic 0. Similarly, in the *Wired-OR* model, the signal net gets the value equal to the logical OR of the two shorted lines. These two models were originally proposed for bipolar technology, and thus not accurate enough for CMOS devices. Bridging faults for CMOS devices are *dominant-AND* and *dominant-OR*. Here, one driver dominates the logic value of the shorted nets, but only for a given logic value.

1.2.4 Delay Fault Model

A *delay fault* causes excessive delay along one or more paths in the circuit. The circuit remains functionally correct, only its delay increases significantly. The most common delay fault model is the *path delay fault*. It considers the cumulative propagation delay of a signal through the path. It is equal to the sum of all gate delays along the path. The major problem with path delay faults is the existence of a large number of paths through a circuit. The number can even be exponential to the number of gates, in the worst case. This makes it impossible to enumerate all path delay faults for test generation and fault simulation. Delay faults require an ordered pair of test vectors $<v_1, v_2>$ to be applied to the circuit inputs. The first vector v_1 sensitizes the path from input to output, while the pattern v_2 creates the transition along the path. Due to the presence of a delay fault, the transition at output gets delayed from its stipulated time. A high-speed, high-precision tester can detect this delay in the transition.

1.3 TEST GENERATION

Test-pattern generation is the task of producing test vectors to ensure high fault coverage for the circuit under test. The problem is commonly referred to as *automatic test-pattern generation* (ATPG). For deterministic testing, test patterns generated by ATPG tools are stored in the *automatic-test equipment* (ATE). During testing, patterns from ATE are applied to the circuit and the responses are collected. ATE compares the responses with the corresponding fault-free ones and accordingly declares the circuit to have passed or failed in testing. The faulty responses can lead the test engineer to predict the fault sources, which in turn may aid in the diagnosis of defects.

Many ATPG algorithms have been proposed in the literature. There are two main tasks in any ATPG algorithm: exciting the target fault and propagating the fault effect to a primary output. A five-valued algebra with possible logic values of $0, 1, X, D,$ and $\overline{D}$ has been proposed for the same. Here, $0, 1,$ and X are the conventional logic values of true, false, and don't care. D represents a composite logic value $1/0$ and $\overline{D}$ represents $0/1$. Logic operations involving D are carried out component-wise. Considering, in this composite notation, logic-1 is represented as $1/1$ and D as $1/0$, "1 AND D" is equal to $1/1$ AND $1/0 = (1$ AND $1)/(1$ AND $0) = 1/0 = D$. Also, "D OR $\overline{D}$" is equal to $1/0$ OR $0/1 = (1$ OR $0)/(0$ OR $1) = 1/1 = 1$. $NOT(D) = NOT(1/0) = NOT(1)/NOT(0) = 0/1 = \overline{D}$.

1.3.1 D Algorithm

This is one of the most well-known ATPG algorithms. As evident from the name of the algorithm, it tries to propagate a D or $\overline{D}$ of the target fault to a primary output. To start with, two sets, *D-frontier* and *J-frontier*, are defined.

D-frontier: This is the set of gates whose output value is X and one or more inputs are at value D or $\overline{D}$. To start with, for a target fault f, D algorithm places a D or $\overline{D}$ at the fault location. All other signals are at X. Thus, initially, *D-frontier* contains all gates that are successors of the line corresponding to the fault f.

J-frontier: It is the set of circuit gates with known output values but not justified yet by the inputs. To detect a fault f, all gates in *J-frontier* need to be justified.

The D algorithm begins by trying to propagate the D (or $\overline{D}$) at the fault site to one of the primary outputs. Accordingly, gates are added to the *D-frontier*. As the D value is propagated, *D-frontier* eventually becomes the gate corresponding to the primary output. After a D or $\overline{D}$ has reached a primary output, justification for gates in *J-frontier* starts. For this, *J-frontier* is advanced backward by placing the predecessors of gates in current *J-frontier* and justifying them. If a conflict occurs in the process, backtracking is invoked to try other alternatives. The process has been enumerated in *Algorithm D-Algorithm* noted next.

Algorithm D-Algorithm

Input: C, the circuit under test.
$\quad\quad$ f, the target fault.
Output: Test pattern if the f is testable, else the declaration "untestable."
Begin
Step 1: Set all circuit lines to X.
Step 2: Set line corresponding to f to D or $\overline{D}$; add it to *D-frontier*.
Step 3: Set *J-frontier* to NULL.
Step 4: Set *pattern_found* to *Recursive_D(C)*.
Step 5: If *pattern_found* then print the primary input values, else print "untestable."
End.

Procedure *Recursive_D(C)*
begin

  If conflict detected at circuit line values or *D-frontier* empty, return false;

  If fault effect not reached any primary output then

      While all gates in *D-frontier* not tried do

      begin

          Let *g* be an untried gate;

          Set all unassigned inputs of *g* to non-controlling values and add them to *J-frontier*;

          pattern_found = *Recursive_D(C)*;

          If *pattern_found* return "TRUE";

      end;

  If *J-frontier* is empty return "TRUE";

  Let *g* be a gate in *J-frontier*;

  While *g* is not justified do

  begin

      Let *k* be an unassigned input of *g*;

      Set *k* to 1 and insert *k* = 1 to *J-frontier*;

      pattern_found = *Recursive_D(C)*;

      If *pattern_found* return "TRUE", else set *k*=0;

  end;

  Return "FALSE";

end.

1.4 DESIGN FOR TESTABILITY (DFT)

To test for the occurrence of a fault at a point in the circuit, two operations are necessary. The first one is to force the logic value at that point to the opposite of the fault. For example, to check whether a gate's output in a circuit is stuck-at 0, it is required to force the gate output to 1. The second task is to make the effect of the fault propagate to at least one of the primary outputs. As noted in Section 1.3, test-pattern generation algorithms essentially do this job. The ease with which a point can be forced to some logic value is known as the *controllability* of the point. Similarly, the effort involved in transferring the fault effect to any primary output is the *observability* measure of the point. Depending upon the complexity of the design, these controllability and observability metrics for the circuit lines may be poor. As a result, the test-generation algorithms may not be able to come up with test sets having high fault coverage. The situation is more difficult for sequential circuits, as setting a sequential circuit to a required internal state may necessitate a large sequence of inputs to be applied to the circuit. *Design for testability* (DFT) techniques attempt to enhance the controllability and observability of circuit lines to aid in the test-generation process. The DFT approaches can be broadly classified into ad hoc and structural approaches.

Ad hoc DFT approach suggests adherence to good design practices. The following are a few examples of the same:

- Inserting test points

- Avoiding asynchronous reset for flip-flops

- Avoiding combinational feedback loops

- Avoiding redundant and asynchronous logic

- Partitioning a big circuit into smaller ones

1.4.1 Scan Design—A Structured DFT Approach

Scan design is the most widely used DFT technique that aims at improving the controllability and observability of flip-flops in a sequential design. The sequential design is converted into a scan design with three distinct modes of operation: *normal mode*, *shift mode*, and *capture mode*. In normal mode, the test signals are deactivated. As a result, the circuit operates in its normal functional mode. In the shift and capture modes, a *test mode* signal is activated to modify the circuit in such a way that test pattern application and response collection becomes easier than the original non-scan circuit.

Figure 1.1a shows a sequential circuit with three flip-flops. For testing some fault in the combinational part of the circuit, the test-pattern generator may need some of the pseudo-primary inputs to be set to some specific values over and above the primary inputs. Thus, it is necessary to put the sequential circuit into some specific state. Starting from the initial state of the sequential circuit, it may be quite cumbersome (if not impossible) to arrive at such a configuration. The structure is modified, as shown in Figure 1.1b, in which each flip-flop is replaced by a *muxed-D scan cell*. Each such cell has inputs like *data input* (DI), *scan input* (SI), *scan enable* (SE), and *clock signal* (CLK). As shown in Figure 1.1b, the flip-flops are put into a chain with three additional chip-level pins: *SI*, *SO*, and *test mode*. To set the pseudo-primary inputs (flip-flops) to some desired value, the signal *test mode* is set to 1. The desired pattern is shifted into the flip-flop chain serially through the line *SI*. If there are k flip-flops in the chain, after k shifts, the pseudo-primary input part of a test pattern gets loaded into the flip-flops. The primary input part of the test pattern is applied to the primary input lines. This operation is known as shifting through the chain. Next, the *test mode* signal is deactivated, and the circuit is made to operate in normal mode. The response of the circuit is captured into the primary and pseudo-primary output lines. The pseudo-primary output bits are latched into the scan flip-flops. Now, the *test mode* signal is activated and the response shifted out through the scan-out pin *SO*. During this scan-out phase, the next test pattern can also be shifted into the scan chain. This overlapping in scan-in and scan-out operations reduces the overall test-application time.

A typical design contains a large number of flip-flops. If all of them are put into a single chain, time needed to load the test patterns through the chain increases significantly. To solve this problem, several alternatives have been proposed:

1. *Multiple Scan Chains*: Instead of a single chain, multiple chains are formed. Separate *SI* and *SO* pin pairs are needed for each such chain.

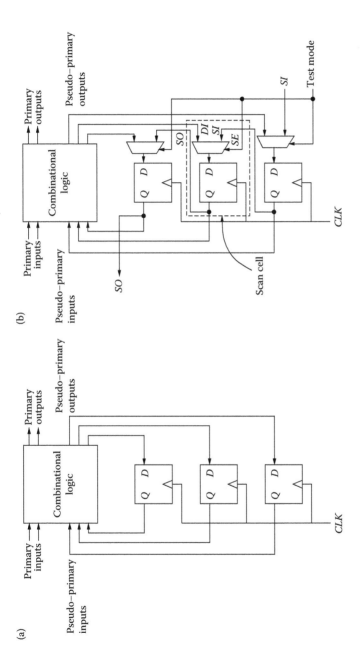

FIGURE 1.1 (a) A sequential circuit, (b) corresponding scan-converted circuit.

2. *Partial Scan Chain*: In this strategy, all flip-flops are not put onto the scan chain. A selected subset of flip-flops, which are otherwise difficult to be controlled and observed, are put in the chain.

3. *Random Access Scan*: In this case, scan cells are organized in a matrix form with associated row- and column-selection logic. Any of the cells can be accessed by mentioning the corresponding row and column numbers. This can significantly reduce unnecessary shifting through the flip-flop chain.

1.4.2 Logic Built-In Self-Test (BIST)

Logic BIST is a DFT technique in which the test-pattern generator and response analyzer become part of the chip itself. Figure 1.2 shows the structure of a typical logic BIST system. In this system, a *test-pattern generator* (TPG) automatically generates test patterns, which are applied to the *circuit under test* (CUT). The *output-response analyzer* (ORA) performs automatic space and time compaction of responses from the CUT into a *signature*. The *BIST controller* provides the BIST control signals, the scan enable signals, and the clock to coordinate complete BIST sessions for the

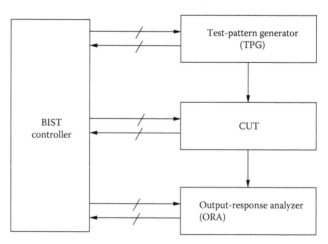

FIGURE 1.2   A typical logic BIST structure.

circuit. At the end of BIST session, the signature produced by the ORA is compared with the *golden signature* (corresponding to the fault-free circuit). If the final space- and time-compacted signature matches the golden signature, the BIST controller indicates a *pass* for the circuit; otherwise it marks a *fail*.

The test-pattern generators (TPGs) for BIST are often constructed from *linear feedback shift registers* (LFSRs). An n-stage LFSR consists of n, D-type flip-flops, and a selected number of XOR gates. The XOR gates are used to formulate the feedback network. The operation of the LFSR is controlled by a *characteristic polynomial $f(x)$* of degree n, given by

$$f(x) = 1 + h_1 x + h_2 x^2 + \cdots + h_{n-1} x^{n-1} + x^n.$$

Here, each h_i is either 1 or 0, identifying the existence or absence of the ith flip-flop output of the LFSR in the feedback network. If m is the smallest positive integer, such that $f(x)$ divides $1 + x^m$, m is called the period of the LFSR. If $m = 2^n - 1$, it is known as a *maximum-length* LFSR and the corresponding characteristic polynomial is a *primitive polynomial*. Starting with a non-zero initial state, an LFSR automatically generates successive patterns guided by its characteristic polynomial. A maximum-length LFSR generates all the non-zero states in a cycle of length $2^n - 1$. Maximum-length LFSRs are commonly used for *pseudo-random* testing. In this, the test patterns applied to the CUT are generated randomly. The pseudo-random nature of LFSRs aids in fault-coverage analysis if the LFSR is seeded with some known initial value and run for a fixed number of clock cycles. The major advantage of using this approach is the ease of pattern generation. However, some circuits show *random-pattern-resistant* (*RP-resistant*) faults, which are difficult to detect via random testing.

The output-response analyzer (ORA) is often designed as a *multiple input signature register* (MISR). This is also a shift register in which, instead of direct connection between two D flip-flops, the output of the previous stage is XORed with one of the CUT outputs and fed to the next D flip-flop in sequence. If the CUT has

m output lines, the number of MISR stages is n, and the number of clock cycles for which the BIST runs is L, then the *aliasing probability* $P(n)$ of the structure is given by

$$P(n) = (2^{(mL-n)} - 1)/(2^{mL} - 1), \quad L > n \geq m \geq 2.$$

If $L \gg n$, then $P(n) \approx 2^{-n}$. Making n large decreases the aliasing probability significantly. In the presence of a fault, the L, m-bit sequence produced at CUT outputs is different from the fault-free response. Thus, if aliasing probability is low, the probability of the faulty signature matching the fault-free golden signature, or the probability of signatures for two or more different faults being the same, is also low.

1.5 POWER DISSIPATION DURING TESTING

Power dissipation in CMOS circuits can be broadly divided into the following components: static, short-circuit, leakage, and dynamic power. The dominant component among all of these is the dynamic power caused by switching of the gate outputs. The dynamic power P_d required to charge and discharge the output capacitance load of a gate is given by

$$P_d = 0.5 V_{dd}^2 \cdot f \cdot C_{load} \cdot N_G,$$

where V_{dd} is the supply voltage, f is the frequency of operation, C_{load} is the load capacitance, and N_G is the total number of gate output transitions ($0 \rightarrow 1$, $1 \rightarrow 0$). From the test engineer's point of view, the parameters such as V_{dd}, f, and C_{load} cannot be modified because these are fixed for a particular technology, design strategy, and speed of operation. The parameter that can be controlled is the switching activity. This is often measured as the *node transition count* (NTC), given by

$$NTC = \sum_{\text{All gates } G} N_G \times C_{load}.$$

However, in a scan environment, computing *NTC* is difficult. For a chain with *m* flip-flops, the computation increases *m*-fold. A simple metric, *weighted transition count* (WTC), correlates well with power dissipation. While scanning-in a test pattern, a transition gets introduced into the scan chain if two successive bits shifted-in are not equal. For a chain of length *m*, if a transition occurs in the first cell at the first shift cycle, the transition passes through the remaining $(m-1)$ cycles also, affecting successive cells. In general, if $V_i(j)$ represents the *j*th bit of vector V_i, the WTC for the corresponding scan-in operation is given by

$$WTC_{scanin}(V_i) = \sum_{j=1}^{m-1} V_i(j) \oplus V_i(j+1)) \cdot (m-j).$$

For the response scan-out operation, shift occurs from the other end of the scan chain. Thus, the corresponding transition count for the scan-out operation is given by

$$WTC_{scanout}(V_i) = \sum_{j=1}^{m-1} V_i(j) \oplus V_i(j+1)) \cdot j.$$

1.5.1 Power Concerns During Testing

Due to tight yield and reliability concerns in deep submicron technology, power constraints are set for the functional operation of the circuit. Excessive power dissipation during the test application which is caused by high switching activity may lead to severe problems, which are noted next.

1. A major part of the power is dissipated as heat. This may lead to destructive testing. At wafer-level testing, special cooling arrangements may be costly. Excessive heat generation also precludes parallel, multisite testing. At board-level, or in-field operation also, overheating may cause the circuit to fail.

Excessive heat dissipation may lead to permanent damage to the chip.

2. Manufacturing yield loss may occur due to power/ground noise and/or voltage drop. Wafer probing is a must to eliminate defective chips. However, in wafer-level testing, power is provided through probes having higher inductance than the power and ground pins of the circuit package, leading to higher power/ground noise. This noise may cause the circuit to malfunction *only* during test, eliminating good unpackaged chips that function correctly under normal conditions. This leads to unnecessary yield loss.

Test power often turns out to be much higher than the functional-mode power consumption of digital circuits. The following are the probable sources of high-power consumption during testing:

1. Low-power digital designs use optimization algorithms, which seek to minimize signal or transition probabilities of circuit nodes using spatial dependencies between them. Transition probabilities of primary inputs are assumed to be known. Thus, design-power optimization relies to a great extent on these temporal and spatial localities. In the functional mode of operation, successive input patterns are often correlated to each other. However, correlation between the successive test patterns generated by ATPG algorithms is often very low. This is because a test pattern is generated for a targeted fault, without any concern about the previous pattern in the test sequence. Low correlation between successive patterns will cause higher switching activity and thus higher-power consumption during testing, compared to the functional mode.

2. For low-power sequential circuit design, states are encoded based on state transition probabilities. States with high transition probability between them are encoded with

minimum Hamming distance between state codes. This reduces transitions during normal operation of the circuit. However, when the sequential circuit is converted to scan-circuit by configuring all flip-flops as scan flip-flops, the scan-cell contents become highly uncorrelated. Moreover, in the functional mode of operation, many of the states may be completely unreachable. However, in test mode, via scan-chain, those states may be reached, which may lead to higher-power consumption.

3. At the system-level, all components may not be needed simultaneously. Power-management techniques rely on this principle to shut down the blocks that are not needed at a given time. However, such an assumption is not valid during test application. To minimize test time, concurrent testing of modules are often followed. In the process, high-power consuming modules may be turned on simultaneously. In functional mode, those modules may never be active together.

Test-power minimization techniques target one or more of the following approaches:

1. Test vector are reordered to increase correlation between successive patterns applied to the circuit.

2. Don't care bits in test patterns are filled up to reduce test power.

3. Logic BIST patterns are filtered to avoid application of patterns that do not augment the fault coverage.

4. Proper scheduling of modules for testing avoids high power consuming modules being active simultaneously.

However, power minimization does not necessarily mean temperature minimization. This is because chip floorplan has a definite role in determining chip temperature. Temperature of a circuit block depends not only upon the power consumption of the block, but also on the power and temperature profiles of surrounding blocks. Scan transitions affect the power consumption of the blocks, but depending upon the floorplan, heating may be affected.

1.6 EFFECTS OF HIGH TEMPERATURE

With continuous technology scaling, the power density of a chip also increases rapidly. Power density in high-end microprocessors almost doubles every two years. Power dissipated at circuit modules are the major source of heat generation in a chip. Power density of high-performance microcontrollers has already reached 50 W/cm^2 at the 100 nm technology node, and it increases further as we scale down the technology. As a result, the average temperature of the die has also been increasing rapidly. The temperature profile of a chip also depends on the relative positions of the power consuming blocks present in the chip. Therefore, physical proximity of high-power consuming blocks in a chip floorplan may result in local thermal hotspots. Furthermore, the three-dimensional (3D) ICs have to deal with the non-uniform thermal resistances (to the heat sink) of the different silicon layers present in the chip. Thus 3D ICs are more prone to the thermal hotspots. High temperature causes the following degradations:

- With an increase in temperature, device mobility decreases. This results in a reduction in transistor speed.

- Threshold voltage of a transistor decreases with an increase in temperature. This results in a reduction in the *static noise margin* (SNM) of the circuit.

- Leakage power increases exponentially with an increase in temperature. This may further increase the chip temperature and result in thermal run-away.

- Resistivity of the metal interconnects increases with an increase in temperature. As a result, the RC delay of the chip increases.

- Reliability of the chip is also a function of chip temperature. According to the Arrhenius equation, *mean time to failure* (MTTF) exponentially decreases with temperature, which is described in Equation 1.1.

$$MTTF = MTTF_0 \exp^{E_a/(k_b \times T)} \tag{1.1}$$

In Equation 1.1, E_a signifies the *activation energy*, k_b is the Boltzmann constant. T represents the temperature and $MTTF_0$ denotes the MTTF at $T = 0°K$.

Thus, high temperature can severely damage the performance of the chip and may result in component failures.

1.7 THERMAL MODEL

Thermal model plays a crucial role in the efficacy of any thermal management technique. A thermal model used in the context of IC design determines the temperature profile of the chip as a function of its geometric dimensions, packaging, power consumptions of the modules present, and the ambient temperature. The thermal model, discussed here, has been adopted from a popular temperature modeling tool, named HotSpot [2]. This tool works on the principle of the *compact thermal model* (CTM). Unlike other numerical thermal-analysis methods (for example, the *finite element method* (FEM)), CTM takes significantly less run time to produce a reasonably accurate prediction of chip temperature. A brief description about the working principle of HotSpot has been presented in the following paragraph.

Power-dissipating devices present in the Si-layer of the chip are the sources of heat generation in the system. A large fraction ($\approx 70\%$) of this generated heat in the Si-layer flows towards the heat sink because of the high thermal conductance values of the layers present in this path. This path is referred to as the primary heat transfer path. Figure 1.3a shows the layers present along the primary heat transfer path of an IC. The *TI*-layer is the *thermal*

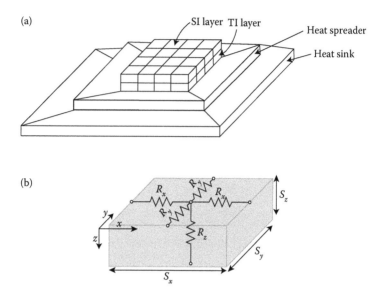

FIGURE 1.3 (a) Different layers of an IC along the primary heat-transfer path, (b) lateral and vertical thermal resistances of an elementary block in a layer.

interface layer present below the Si-layer to enhance the thermal coupling between the Si-layer and the *heat spreader*. Each layer is further divided into several identical (except the peripheral blocks) elementary blocks. The smaller the size of an elementary block, the higher the accuracy and the run time of the thermal model. Power consumption of an elementary block is determined using the power density values of the power-consuming blocks placed in it and their respective areas of overlap with the elementary block. In Figure 1.3a, the Si-layer is divided into sixteen elementary blocks using a 4 × 4 grid mesh. In HotSpot, if the Si-layer contains b number of blocks within it, all the layers below the Si-layer should also be divided into a similar b number of blocks. As shown in Figure 1.3a, along with this 4 × b number of blocks, the heat spreader layer contains four extra peripheral blocks and the heat sink layer contains eight extra peripheral blocks. Therefore, the total number of blocks present in the chip (*Num_Block*) is $4 \times b + 12$. A CTM works on the duality between the thermal and the electrical quantities. Following this duality principle, HotSpot replaces all the elementary blocks in the chip by electrical nodes in their position, containing equivalent resistances and current sources connected to it. Electrical resistance connected to a node is considered to be equivalent to the thermal resistance of that block. The magnitude of the current source connected to a node is set according to the power consumption of the circuit module holding that block. Figure 1.3b shows different horizontal and vertical thermal resistances considered inside a block. Equations 1.2 and 1.3 represent the horizontal thermal resistances, R_x and R_y, along the x- and y-directions, respectively. Equation 1.4 represents the vertical thermal resistance, R_z, along the z-direction. The unit of the thermal resistance is taken to be $K/Watt$.

$$R_x = (1/K_{layer})(0.5 \times S_x /(S_y \times S_z)) \tag{1.2}$$

$$R_y = (1/K_{layer})(0.5 \times S_y /(S_z \times S_x)) \tag{1.3}$$

$$R_z = (1/K_{layer})(0.5 \times S_z /(S_x \times S_y)) \tag{1.4}$$

S_x, S_y, and S_z represent the lengths of the block along the x-, y-, and z-directions, respectively. K_{layer} represents the thermal conductivity of a particular layer. Integrating the equivalent electrical circuit of each of the individual blocks, the total equivalent electrical circuit of the chip is extracted. In order to get the temperature values for all the blocks, this equivalent electrical circuit is solved and the node voltages are determined. Inputs to the thermal model are: geometrical information of the chip, thermal conductivities of different layers present in it, dimensions and power consumption values of the modules, and the ambient temperature. Using the above information, the conductance matrix $[Cond]_{Num\_Block \times Num\_Block}$ and the power matrix $[Pow]_{Num\_Block \times 1}$ ($Num\_Block = 4 \times b + 12$) for the equivalent electrical circuit are calculated. For a particular floorplan, temperatures of the blocks can be calculated by solving Equation 1.5.

$$[Cond]_{Num\_Block \times Num\_Block} \times [Temp]_{Num\_Block \times 1} = [Pow]_{Num\_Block \times 1} \quad (1.5)$$

$[Temp]_{Num\_Block \times 1}$ represents the temperature matrix, containing the temperature values for each block in the chip. Equation 1.5 is solved using the LU-decomposition method.

1.8 SUMMARY

This chapter has presented an introduction to the VLSI testing process [3]. The role of testing in ensuring the quality of the manufactured ICs has been enumerated. Different fault models have been illustrated. Design for test techniques have been discussed to enhance the testability of designs. Test-generation algorithms have been enumerated. This has been followed by power and thermal issues in the testing process. Sources of power dissipation, heat generation, and a compact thermal model for estimating chip temperature have been presented.

REFERENCES

1. G. Moore, "Cramming More Components onto Integrated Circuits", *Electronics*, vol. 38, No. 8, pp. 114–117, 1965.
2. K. Skadron, M. Stan, W. Huang, S. Velusamy, K. Sankaranarayanan, D. Tarjan, "*Temperature-aware Microarchitecture: Extended Discussion and Results,*" Univ. of Virginia Dept. of Computer Science Tech. Report CS-2003-08, Tech. Rep., 2003.
3. L.T. Wang, C.W. Wu, X. Wen, "*VLSI Test Principles and Architectures*", Morgan Kaufmann, 2006.

CHAPTER 2

Circuit-Level Testing

2.1 INTRODUCTION

A digital VLSI circuit consists of logic elements, such as gates and flip-flops. Test strategies for these circuits consist of applying a set of test patterns, noting their responses, and comparing these responses with the corresponding fault-free values. Depending upon the fault model, a single pattern or a sequence is entrusted with the task of detecting one or more faults. Test patterns are generated by software tools, commonly known as *automatic test-pattern generation* (ATPG). The generated test set can often be found to possess the following properties:

1. A large number of bits in the test set are left uninitialized by the ATPG. Values of these bits are not significant in detecting faults by a particular pattern. These are commonly known as *don't cares*. These don't-care bits can be utilized in different ways: reducing test-set size via compression, reducing test power and associated heating by circuit transition reduction, and so on.

2. Sequential circuits are often converted into scan circuits, in which all flip-flops of the circuit are put onto a chain. A

dedicated pin (called *scan-in*) is used to input patterns into these flip-flops serially. The contents of these flip-flops are made visible to the outside by shifting them out serially through another dedicated output pin (called *scan-out*). Introduction of such a scan chain allows the test engineer to consider any sequential circuit as a combinational one and apply combinational ATPG algorithms to generate test patterns for the circuit. However, test-application process becomes longer, as test and response bits are shifted serially through the flip-flop chain. It also creates unnecessary ripples in the chain, which, in turn, cause transitions in the circuit nodes, leading to power consumption and associated heat generation. Test power, as well as thermal optimization, is possible by reordering the test patterns and/or designing the scan chains efficiently.

3. ATPG algorithms often produce more than one pattern per fault. The algorithms may be instructed to stop considering a fault for test-pattern generation (fault dropping) after a fixed number n (say, $n = 2, 3$, etc.) of patterns have been generated (n-detect test set) targeting that fault. It leaves the scope of choosing the final patterns from the set of redundant patterns generated by the algorithm. The selected pattern set may have good power and thermal implications.

Apart from the test set, power consumption and heat generation can also be controlled by designing the scan chains carefully. For example, instead of putting all flip-flops on a single long chain, consider multiple, shorter-length chains. This can restrict the scan ripples from traveling a long distance, creating large number of toggles at the combinational circuit inputs. Flip-flop types may also determine the toggle rate. In normal circuits, generally D-type flip-flops are used as sequential elements. However, using other types of flip-flops in the chain may reduce circuit transitions. Thus, designing chains such that, in test mode, some of the flip-flops get converted to some desired type (rather than D) could be an effective approach toward thermal-aware testing.

The *built-in self-test* (BIST) approach of testing removes the requirements for external tester and test-generation algorithms. On-chip pseudo-random sequence generators are used to generate test patterns and apply to the circuit under test. However, BIST generates a large number of patterns that do not detect new faults. Thus, making BIST sessions power- and thermal-aware becomes a challenge.

In the remaining part of the chapter, these temperature-aware testing methods have been discussed in detail. Section 2.2 discusses pattern reordering techniques. Section 2.3 covers a thermal-aware don't care bit filling strategy. Section 2.4 presents scan-cell optimization. Section 2.5 discusses thermal-aware BIST design, while Section 2.6 summarizes the chapter.

2.2 TEST-VECTOR REORDERING

Vector reordering is an effective tool to reduce circuit transitions during testing. Many of the fault models, particularly those targeting stuck-at faults, are insensitive to the order in which patterns are applied to the circuit. This is especially true for pure combinational and scan-converted sequential circuits. Given a set of test patterns $T = \{v_1, v_2, \dots v_n\}$, vector reordering gives a permutation of these test patterns identifying the order in which those are applied to the circuit under test. Different orders will excite parts of the circuit to different magnitudes, affecting power consumption and heating of circuit blocks. To identify a good order, a Hamming distance-based heuristic technique is first presented to reduce the temperature of the hottest block identified by the thermal simulation corresponding to the initial ordering of patterns. This is followed by a meta-heuristic technique developed around *particle swarm optimization* (PSO) to identify a good ordering. Experimental results are presented for all the techniques and compared with a power-aware reordering of patterns [1].

2.2.1 Hamming Distance-Based Reordering

This is a highly greedy approach to reduce the peak temperature of circuit blocks. A circuit C is assumed to be consisting of m blocks, $\{b_1, b_2, \ldots b_m\}$. Division of C into blocks can be done taking into view the granularity used for thermal simulation. A coarse grained blocking (involving more circuit elements in a block) will have fewer blocks with faster but less accurate simulation. A finer granularity, on the other hand will require more time for thermal simulation. Let $T = \{v_1, v_2, \ldots v_n\}$ be the set of test vectors. The initial ordering O_{orig} is considered to be the sequence $<v_1, v_2, \ldots v_n>$. For this sequence, power consumed by individual blocks $b_1, b_2, \ldots b_m$ is calculated. A thermal simulation is performed to get the temperature values of all the blocks. The maximum among all of these values is noted. Every block is assigned a weight factor equal to the ratio of its temperature to this maximum. Thus, a weight value close to 1 indicates that the corresponding block is one of those having a temperature close to the peak for the chip. Such a block should get more attention, in order to reduce the peak temperature of the chip, compared with a block having a relatively low weight factor (indicating a relatively cool block). Now, temperature of a block can be controlled by controlling the power consumption in its constituent logic elements. A lesser switching activity in the subset of primary inputs of the circuit that eventually affect the logic gates of the block may have a significant effect on reducing the power consumed by those gates. Keeping this observation in view, the optimization process first identifies the primary inputs of the circuit belonging to the cone of dependency of the blocks. A typical example circuit divided into blocks along with their primary input cones has been shown in Figure 2.1.

Pattern v_1 is taken as the first one in the reordered pattern set. To select the next pattern to follow v_1, cost for each remaining pattern is calculated. Specifically, the activity corresponding to vector v_i, $Activity_i$, is calculated as follows. For each block b_j, let v_{ij} be the part of the pattern considering the inputs in the primary

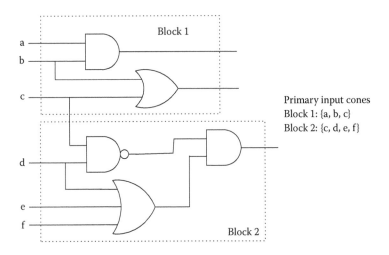

FIGURE 2.1   Blocks in an example circuit with corresponding cones of dependency.

input cone of b_j alone. Hamming distance between v_{1j} *and* v_{ij} is calculated and multiplied by *Weight$_j$*. Let this product be called *Activity$_{ij}$*. Thus,

$$Activity_{ij} = Hamming\_distance\,(v_{1j},\ v_{ij}) \times Weight_j.$$

All *Activity$_{ij}$* values are summed to get *Activity$_i$*.

$$Activity_i = \sum_{j=1}^{m} Activity_{ij}$$

In this way, all activities, *Activity$_2$*, *Activity$_3$*, ... *Activity$_n$*, are computed. The minimum among all of these corresponds to the pattern that, if followed by v_1, will excite hotter blocks proportionally less. If this minimum corresponds to pattern v_k, it is removed from the set of patterns O_{orig} and is taken to follow v_1 in the reordered pattern set. Pattern v_k now assumes the role of v_1, and the algorithm continues to find the next candidate to follow v_k.

The process stops when O_{orig} becomes empty. The sequence O_{final} holds the final reordered test set. The whole procedure is presented in the *Algorithm Hamming_Reorder*.

Algorithm Hamming_Reorder

Input: Set of test vectors $\{v_1, v_2, \ldots v_n\}$, Circuit consisting of blocks $\{b_1, b_2, \ldots b_m\}$.
Output: Reordered test vectors in sequence O_{final}.
Begin
Step 1: Set initial order $O_{orig} = <v_1, v_2, \ldots v_n>$.
Step 2: Compute block-level power values corresponding to sequence O_{orig}
Step 3: Perform thermal simulation to get temperature of blocks b_1 through b_m.
Step 4: Set *Max_Temp* = Temperature of the hottest block among b_1 through b_m.
Step 5: For each block $b_j \in \{b_1, b_2, \ldots b_m\}$ do
　　　　Set PI_Cone_j = Set of primary inputs affecting logic elements in b_j.
　　　　Set $Weight_j$ = (Temperature of b_j) / Max_Temp.
Step 6: Set $O_{final} = <v_1>$; Remove v_1 from O_{orig}
Step 7: While $O_{orig} \neq \varphi$ do
　　begin
　　Let v_t be the last vector in O_{final}.
　　For each vector v_k in O_{orig} do
　　　　$Activity_k = \sum_{j=1}^{m}$(Hamming distance between parts of v_t and v_k affecting block b_j) × $Weight_j$.
　　Select vector v_m with minimum $Activity_m$.
　　Append v_m to O_{final}.
　　Remove v_m from O_{orig}
　　end;
Step 8: Output O_{final}.
End.

TABLE 2.1 Temperature Reduction in Hamming
Distance-Based Approach

Circuit	Peak Temperature in Degrees K		% Reduction in Peak Temperature
	Original	Hamming	
s5378	380.89	362.81	4.75
s9234	393.82	374.95	4.79
s13207	408.15	399.95	2.01
s15850	419.58	410.13	2.25
s38417	453.25	441.84	2.52
s38584	491.85	475.95	3.23

In Table 2.1, the peak temperature values in degrees Kelvin have been reported for a number of benchmark circuits corresponding to test patterns generated by the Synopsys Tetramax ATPG. Temperatures have been reported for the original unordered-test set and the test set obtained via reordering using the Hamming distance minimization approach presented in this section. The percentage reductions have also been noted. It can be observed that the Hamming distance-based reordering approach could reduce the peak temperature in the range of 2% to 4.7%. The scheme is computationally simple as it takes very little CPU time to run the program.

2.2.2 Particle Swarm Optimization-Based Reordering

A major shortcoming of the greedy heuristic approach presented in Section 2.2.1 is that it does not reevaluate the peak temperature behavior of blocks once some of the vectors have changed positions in the ordering. A more elaborate search is expected to reveal more potential orders. In this section, a *particle swarm optimization* (PSO) based approach has been presented for the reordering problem which often comes with better solutions.

PSO is a population-based stochastic technique developed by Eberhart and Kennedy in 1995. In a PSO system, multiple candidate solutions coexist and collaborate between themselves. Each solution, called a *particle*, is a position vector in the search space. The position of a particle gets updated through generations according to its own experience as well as the experience of the current best particle. Each particle has a fitness value. The quality of a particle is evaluated by its fitness. In every iteration, the evolution of a particle is guided by two best values—the first one is called the *local best* (denoted as *pbest*) and the second one the *global best* (denoted as *gbest*). The local best of a particle is the position vector of the best solution that the particle has achieved so far in the evolution process across the generations, the value being stored along with the particle. On the other hand, the global best of a generation corresponds to the position vector of the particle with the best fitness value in the current generation. After finding these two best particle structures, the velocity and position of a particle are updated according to the following two equations.

$$v_{m+1}^i = wv_m^i + c_1 r_1 (pbest^i - x_m^i) + c_2 r_2 (gbest_m - x_m^i) \quad (2.1)$$

$$x_{m+1}^i = x_m^i + v_{m+1}^i \quad (2.2)$$

Here, v_{m+1}^i is the velocity of the particle i at the $(m + 1)$th iteration, x_m^i is the current position (solution) of the particle, r_1 and r_2 are two random numbers in the range 0 to 1, and c_1 and c_2 are two positive constants. The constant c_1 is called the self-confidence (cognitive) factor, while c_2 is the swarm-confidence (social) factor. The constant w is called the inertia factor. The first term in Equation 2.1 presents the effect of inertia on the particle. The second term in the equation represents particle memory influence, while the third term represents swarm influence. The inertia weight w guides the global and local exploration. A larger inertia weight puts emphasis on global exploration, while a smaller inertia weight tends to entertain local exploration to fine-tune the current search area. The second equation, Equation 2.2, corresponds to the updating of the particle through its new velocity.

Inspired by success in solving problems in a continuous domain, several researchers have attempted to apply PSO in a discrete domain as well. In a discrete PSO (DPSO) formulation, let the position of a particle (in an n dimensional space) at the kth iteration be $p_k = <p_{k,1}, p_{k,2}, \ldots p_{k,n}>$. For the ith particle, the quantity is denoted as p_k^i. The new position of particle i is calculated as follows.

$$p_{k+1}^i = (w * I \oplus c_1 r_1 * (p_k \rightarrow pbest^i) \oplus c_2 r_2 * (p_k \rightarrow gbest_k) \quad (2.3)$$

In Equation 2.3, $a \rightarrow b$ represents the minimum-length sequence of swapping to be applied on components of a to transform it to b. For example, if $a = <1, 3, 4, 2>$ and $b = <2, 1, 3, 4>$, $a \rightarrow b = <swap(1, 4), swap(2, 4), swap(3, 4)>$. The operator $\oplus$ is the fusion operator. Applied on two swap sequences a and b, $a \oplus b$ is equal to the sequence in which the sequence of swaps in a is followed by the sequence of swaps in b. The constants w, c_1, c_2, r_1, and r_2 have their usual meanings. The quantity $c * (a \rightarrow b)$ means that the swaps in the sequence $a \rightarrow b$ will be applied with a probability c. I is the sequence of identity swaps, such as $<swap(1, 1), swap(2, 2), \ldots swap(n, n)>$. It corresponds to the inertia of the particle to maintain its current configuration. The final swap corresponding to $w * I \oplus c_1 r_1 * (p_k \rightarrow pbest^i) \oplus c_2 r_2 * (p_k \rightarrow gbest_k)$ is applied on particle p_k^i to get particle p_{k+1}^i. In the following, a DPSO-based formulation has been presented for the reordering problem.

Particle Structure: If n is the number of test vectors under consideration, a particle is a permutation of numbers from 1 to n. The particle identifies the sequence in which patterns are applied to the circuit during testing.

Initial Population Generation: To start the PSO evolution process, the first generation particles need to be created. For the vector ordering problem, the particles can be created randomly. If the population size is x, x different permutations of numbers from 1 to n are generated randomly.

Fitness Function: To estimate the quality of the solution produced by any particle, ideally, a thermal simulator must be used. By applying the test patterns in the sequence suggested by the particle, the power consumed by different circuit blocks is obtained. These power values, along with the chip floorplan are fed to a thermal simulator to get the temperature distribution of circuit blocks. However, the approach is costly, particularly for application in evolutionary approaches such as PSO, as the fitness computation must be repeated for all particles in each generation. Instead, in the following, a transition count-based metric is presented that can be utilized to compare the thermal performances of particles.

When a sequence of test patterns is applied to a circuit, each circuit block sees a number of transitions. These transitions lead to proportional consumption of dynamic power by the block. Because power consumption has a direct impact on the thermal behavior of the block, the transition count affects the temperature of the block. However, the temperature of a block is also dependent on the thermal behavior of its neighbors. This needs to be taken into consideration when formulating the fitness function.

First, the transitions in each block corresponding to the initial test-vector sequence are calculated. For block b_i, let the quantity be $T_{initial}^i$. The block facing the maximum number of transitions is identified. Let the corresponding transition count be T_{max}. To block b_i, a weight factor W_i is assigned, given by $W_i = T_{initial}^i / T_{max}$. Thus, W_i is equal to 1 for the block(s) seeing the maximum number of transitions with the initial sequence of test patterns. Because temperature also depends upon the floorplan, next, the neighbors are identified for each block. A neighbor of a block shares at least a part of its boundary with this block. If $\{b_{i1}, b_{i2}, \dots b_{iN}\}$ is the set of neighbors of b_i, the average weight of neighboring blocks is calculated as

$$W_{avg}^i = \frac{1}{N} \sum_{j=i1}^{iN} W_j.$$

Criticality of block b_i, CR_i, toward thermal optimization is defined as

$$CR_i = \frac{1}{1 + |W_i - W_{avg}^i|} \tag{2.4}$$

From Equation 2.4, it can be observed that *Criticality* of a block is a measure of thermal gradient between the block and its neighbors. A higher value of CR_i indicates that the neighboring blocks are also facing transitions similar to the block b_i. As a result, neighbors will possibly have a temperature close to that of b_i. The possibility of heat transfer to the neighbors is less. If CR_i is low, the surrounding blocks of b_i are seeing either a higher or lower number of transitions, on average, compared to b_i. Hence, if b_i is experiencing a high numbers of transitions, surrounding blocks are likely to be cooler, enabling the transfer of heat generated at b_i to them, thus reducing the temperature of b_i, in turn.

To assign a fitness value to particle p, first the transitions are counted for each circuit block, due to the application of patterns in the sequence specified by p. Let the number of transitions of block b_i corresponding to the pattern sequence p be T_p^i. The difference between T_p^i and $T_{initial}^i$ (corresponding to the initial ordering) is calculated. This difference is multiplied by the weight factor W_i and criticality CR_i of block b_i. The sum of these products across all the blocks of the circuit forms the fitness of p. Thus, the fitness of p is given by

$$Fitness(p) = \sum_{\text{All blocks } i} W_i \times CR_i \times (T_p^i - T_{initial}^i) \tag{2.5}$$

A particle with a low value of this fitness is expected to be of good quality. As can be observed from Equation 2.5, the fitness function encourages particles to reduce the transitions for blocks with higher weight (W_i) and higher criticality (CR_i) values. Such a block is one facing a large number of transitions in the initial ordering, while its surrounding blocks are also having a similar number of transitions.

Evolution of Generations: Once created in the initial population, particles evolve over generations to discover better solutions. As noted in Equation 2.3, at each generation, a particle aligns with the local and global bests with some probabilities. This alignment is effected through *swap* operations. Consider a particle $p = \{v_1, v_2, v_3, v_4, v_5, v_6, v_7\}$ corresponding to a test pattern set with seven patterns in it. This particle suggests that the pattern v_1 be applied first, followed by v_2, v_3, ... v_7 in sequence. A swap operator *swap(x, y)* applied on particle p will interchange the patterns at positions x and y in p to generate a new particle. For example, *swap(2, 4)* converts p to $\{v_1, v_4, v_3, v_2, v_5, v_6, v_7\}$. A sequence of such swap operations are needed to align a particle to its local and global bests. For example, to align a particle $p = \{v_1, v_4, v_3, v_5, v_2, v_7, v_6\}$ with its corresponding local best, say $\{v_2, v_4, v_5, v_7, v_1, v_6, v_3\}$, the swap sequence is $\{swap(1, 5), swap(3, 4), swap(4, 6), swap(6, 7)\}$. Once the swap sequences to align a particle with its local and global bests have been identified, individual swaps are applied with some probabilities. For the local best, it is $c_1 r_1$, and for the global best, it is $c_2 r_2$.

Particles evolve over generations through the sequence of swap operators. To determine the termination of the evolution process, two terminating conditions can be utilized as noted next.

1. PSO has already evolved for a predefined maximum number of generations.

2. The best fitness value of particles has not improved further for a predefined number of generations.

The overall DPSO formulation for the test-vector reordering problem has been noted in *Algorithm DPSO_Reorder*.

Algorithm DPSO_Reorder

Input: A set of n test vectors with an initial ordering.
Output: Reordered test-vector set.
Begin
Step 1: For $i = 1$ to *population_size* do   // Create initial population.
    begin

        particle[i] = A random permutation of numbers 1 to n;
        Evaluate *Fitness*[i] using Equation 2.5;
        pbest[i] = *particle*[i];

    end;
Step 2: *gbest* = particle with best fitness value;
    generations = 1;
    gen_wo_improv = 0;
Step 3: While *generations* $<MAX\_GEN$ and *gen_wo_improv* $<MAX\_GEN\_WO\_IMPROV$ do
    begin   // Evolution through generations

        old_gbest = gbest;
        For $i = 1$ to *population_size* do
        begin

            $Swap\_seq_{local}$ = Swap sequence to align *particle*[i] with *pbest*[i];
            $Swap\_seq_{global}$ = Swap sequence to align *particle*[i] with *gbest*;
            Generate two random numbers r_1 and r_2 in the range (0,1);
            particle[i] = Resulting particle via application of $Swap\_seq_{local}$ and $Swap\_seq_{global}$ to *particle*[i] with probabilities $c_1 r_1$ and $c_2 r_2$;
            Evaluate *Fitness*[i] using Equation 2.5;
            If *Fitness*[i] $<$ fitness of *pbest*[i], set *pbest*[i] = *particle*[i];
            If *Fitness*[i] $<$ fitness of *gbest*, set *gbest* = *particle*[i];

> **end;**
> If (gbest = old_gbest) then *gen_wo_improv++;*
> *generations ++;*

end;

Step 4: Output gbest;

End.

Table 2.2 reports the peak temperature values in degrees Kelvin under different ordering policies for a number of benchmark circuits. The first column notes the name of the circuit. The second column reports the peak temperature values with the input ordering of test vectors. The column "Hamming ordering" corresponds to the Hamming distance-based reordering policy suggested in Section 2.2.1. The column marked "Peak power ordering" corresponds to a DPSO implementation of the reordering policy in which, as the fitness measure, only the peak power consumed by circuit blocks have been considered. The optimization algorithm thus attempts to minimize this peak power consumption. The resulting peak temperature values have been reported in the table. The column marked "DPSO" contains results corresponding to the DPSO formulation presented in this section. The last column is another modified version of the DPSO algorithm. Here, instead of modeling the fitness of a particle

TABLE 2.2 Temperature Reduction in DPSO-Based Approach

	Peak Temperature in Degrees Kelvin				
Circuit	Original Ordering	Hamming Ordering	Peak Power Ordering	DPSO	DPSO with Thermal Simulator
s5378	380.89	362.81	359.30	354.27	350.95
s9234	393.82	374.95	368.67	365.91	357.02
s13207	408.15	399.95	391.11	383.34	381.16
s15850	419.58	410.13	413.86	392.50	389.01
s38417	453.25	441.84	446.50	436.25	433.46
s38584	491.85	475.95	477.28	464.33	461.32
Average % reduction		3.26	3.69	5.99	6.95

with Equation 2.5, a thermal simulator has been integrated into the formulation. A particle gives a sequence for application of test patterns. This results in dynamic power consumption. This power profile and chip floorplan are fed to the thermal simulator to get the temperature profile. From this, the peak temperature is determined. The direct incorporation of thermal simulator into the DPSO algorithm is expected to result in better peak temperature optimization than through the modeling in Equation 2.5. In the table also it can be noted that this approach results in the maximum reduction in peak temperature. The DPSO formulation using Equation 2.5 also performs quite well, and as reported in the table, is inferior to the thermal simulator integrated approach by less than 1%, on average. However, the time needed for the DPSO approach integrated with the thermal simulator is quite high. It has been observed that it takes about four times greater CPU time than the DPSO approach with fitness modeled by Equation 2.5. The other two approaches, that is, the Hamming distance-based one and the peak power minimization approach perform quite poorly, compared with DPSO.

2.3 DON'T CARE FILLING

Test patterns generated by the *automatic test-pattern generation* (ATPG) tools leave a large number of bits in the patterns unspecified. These unspecified bits are called *don't cares*, as their values do not affect the fault coverage of the test-pattern set. It is not unlikely to have up to about 80% to 90% of bits of a test-pattern set as don't cares. Test engineers have the liberty to fill up these bits to optimize several test parameters, such as test compression, test-power reduction, etc. These bits can also be customized to control the temperature of the chip during test.

To ease the test-generation process, all flip-flops in a circuit are configured as scan flip-flops in the test mode. The flip-flops are put on a single or multiple serial chain(s). Test patterns are shifted serially through the chain(s) in the scan mode. Similarly, test responses are shifted out serially. When a pattern shifts through such a chain, the constituent flip-flops toggle a good number of times before settling to the final test pattern. For example, in a chain with four flip-flops, if the final test pattern is "0101," assuming that the initial content of the scan chain is all zeros, the intermediary values are as follows.

Sequence	FF1	FF2	FF3	FF4
Initial:	0	0	0	0
1st shift:	1	0	0	0
2nd shift:	0	1	0	0
3rd shift:	1	0	1	0
4th shift:	0	1	0	1

In these four shifts, a total of ten transitions are seen by the flip-flops. Transitions in these flip-flops, in turn, create transitions in the combinational logic fed by them, leading to power consumption in the corresponding blocks. Test-power reduction aims at reducing these transitions during the scan-in process. Don't cares in test patterns are often filled in a fashion so that successive bits in pattern become similar, reducing transitions in the scan flip-flops. Popular ways to fill the don't care bits are as follows:

- *Random fill*: Fills the don't cares randomly.

- *Zero fill*: Fills the don't cares with zeros.

- *One fill*: Fills the don't cares with ones.

- *Minimum transition fill*: Fills a block of don't cares with the bit preceding or following it.

For thermal-aware filling, it is also necessary to consider the floorplan of the chip. Transitions in two different flip-flops in a block may have different effects on the power consumption of the block. In the following section, a power and thermal estimator has been presented based on the transitions in the flip-flops affecting the block.

2.3.1 Power and Thermal Estimation

For estimating the power consumed by a block when a set of test patterns is applied to the circuit, the first step is to identify the flip-flops having critical impact on the block. A circuit block may contain some flip-flops. These are included in the set of critical flip-flops for the block. A block also gets input from its neighboring blocks. A flip-flop in a neighboring block may have a path to one of the inputs to this block. If the path contains combinational logic only, the flip-flop in the neighboring block is also taken to be a critical flip-flop of the current block.

Once the critical flip-flops for all blocks have been identified, a power simulator is utilized to get the power consumption of the circuit elements by applying ten thousand random patterns. In the process, the power consumed by individual blocks and flip-flops are identified. For block b_i, if P_i^{random} is the power consumed by the block and C_i is the total power consumed by its critical flip-flops, a weight W_i is assigned to the block, given by $W_i = P_i^{random}/C_i$. Thus, a higher weight of a block indicates that the block will consume more power when transitions occur in its critical flip-flops.

Now, the power (mainly, dynamic power) consumed by a flip-flop is directly proportional to the number of transitions occurring in it. For a given test set, when the patterns are shifted in through the scan chain, transitions occur in the flip-flops. By counting the transitions in the critical flip-flops, it is possible to arrive at a relative power consumption measure of circuit blocks. For a test set, if the critical flip-flops of block b_i see a total of T_i transitions, the relative power consumption of the block can be estimated as

$$P_i = T_i \times W_i.$$

If, for a cluster of blocks, their relative power estimates are high, it is expected that a thermal hotspot will get created there. For block b_i, let it have N neighbors with the total relative power of all neighbors being P_i^{ne}. *Criticality* of block b_i for thermal optimization is then defined as

$$CR_i = (P_i + P_i^{ne})/(N + 1).$$

If the criticality of a block is high, it is expected that the block will have high temperature. Thus, the don't care filling policy should try more to reduce power consumption of this block and its neighbors, compared with a block with low criticality.

2.3.2 Flip-Select Filling

This section presents a heuristic approach [2] to fill the don't care bits of a test set. It attempts to carry out the activity aiming at minimizing the maximum criticality of blocks in the circuit. Thus, for a circuit with N blocks, the objective is

$$Minimize\,(Maximum_{1 \le i \le N}(CR_i)).$$

The approach starts by filling the don't cares in the test set in four different ways: random-fill, zero-fill, one-fill, and minimum-transition-fill. For each of the resulting four filled test sets, thermal simulation is done with a thermal simulator. The test set requiring minimum peak temperature is chosen to be the initial test set on which the core flip-select algorithm works as follows. For the selected test set, let *orig_c* be the maximum CR_i among all blocks b_i ($1 \le i \le N$) of the circuit. It may be noted that in this set, in each pattern, the don't care bits are filled up. A pattern is picked up. For this pattern, the filled don't care bits are flipped, one at a time, to their complemented values. After flipping a bit, the maximum criticality values of all blocks are reevaluated. Let the new value be *temp_c*. If *temp_c* is less than *orig_c*, the corresponding flip is retained and *orig_c* is updated to *temp_c*. Otherwise, the flipped bit is switched back to its original filled value. The process is repeated for every test pattern. This procedure of don't care filling has been noted in Algorithm *Flip_Select*.

Algorithm Flip_Select

Input: Circuit under test C, incompletely specified test set T
Output: Completely specified thermal-aware test set
Begin
Step 1: Fill-up don't cares in T using (i) random-fill, (ii) zero-fill, (iii) one-fill, (iv) minimum-transition-fill;
Step 2: Perform thermal simulation of C for each of the resulting filled test sets in Step 1;
Step 3: Set *Init_set* to the test set from Step 1 resulting in the minimum peak temperature;
Step 4: Set *orig_c* to the maximum criticality of blocks in C for *Init_set*;
Step 5: For each test pattern $t \in T$ do
    begin
        For each don't care bit in t do
        begin
            Flip the corresponding bit of *Init_set*;
            Set *temp_c* to the maximum criticality of blocks in C for this *Init_set*;
            If (*temp_c* < *orig_c*) then
                Set *orig_c* to *temp_c*;
            Else
                Flip the bit back to its value in *Init_set*;
        end;
    end;
Step 6: Output the modified fully specified set *Init_set*;
End.

It may be noted that the order in which the test patterns are picked up in Step 5 of the *Flip_Select* algorithm can have a significant effect on transitions in the blocks. This is particularly true because a pattern is picked up only once for choosing the values of don't care bits. Though different orderings can be tried out, the order in which a pattern with more don't care bits is given preference has generally been found to produce better results.

TABLE 2.3   Peak Temperature for Different Filling Techniques

	Peak Block Temperature in Degrees Kelvin					
					Flip_Select	
Circuit	Zero-Fill	One-Fill	Min.-Tran-Fill	Random-Fill	Unordered	Ordered
s5378	358.49	367.03	334.21	486.51	319.72	318.09
s9234	376.98	429.45	400.99	495.39	355.02	330.22
s13207	359.62	397.13	388.01	508.27	347.29	329.44
s15850	374.51	415.53	391.72	501.13	356.67	325.92
s38417	381.37	396.37	377.99	517.15	354.95	320.97
s38584	386.17	432.54	410.14	520.60	353.97	321.42
b14	353.36	367.42	358.69	492.24	344.67	343.26
b15	364.68	374.94	365.04	489.00	359.00	359.98
b20	367.89	383.94	371.66	497.92	357.04	356.14
b21	374.91	392.91	383.76	510.56	362.81	360.45
b22	382.32	399.51	385.80	519.08	371.50	344.68
Avg. % Imp.	4.83, 8.97	10.70, 14.50	6.70, 10.70	29.88, 32.95		

Table 2.3 reports the results of experimentation on different don't care filling techniques for a number of benchmark circuits. Apart from the traditional filling techniques such as zero-fill, one-fill, minimum-transition-fill, and random-fill, the *Flip_Select* results under two circumstances have been noted. The "Unordered" case reports the results in which patterns have been picked up in Step 5 of the algorithm as per their occurrence in the test set. The column "Ordered" corresponds to the case in which the patterns have been ordered in the decreasing number of don't care bits in them. The last row notes the average improvements in peak temperature values in *Flip_Select* (both unordered and ordered cases) over the other four heuristics.

2.4 SCAN-CELL OPTIMIZATION

In addition to reordering and customization of don't care bits in test vectors, scan chain architecture of a circuit can also be modified to achieve peak temperature minimization. In traditional scan-based test architectures, scan chains are designed using D flip-flops.

Additional multiplexers are incorporated to control the inputs of these flip-flops. In the functional mode of operation of the circuit, inputs from lines in the circuit reach these flip-flops. In test mode, test inputs reach these flip-flops from the scan-in pin of the chip. Incorporating some additional logic, types of these flip-flops could be controlled to be either *D*- or *T*-type in test mode. As shown in Figure 2.2, in the normal mode of operation, *Test_mode* is set to 0. As a result, *Normal_data* from the circuit internal line reaches the flip-flop and gets stored in it. In test mode, the control *Test_mode* is set to 1. Now, if *DT_control* is set to 0, the value sent through Scan_data gets stored into the flip-flop. On the other hand, if *DT_control* is set to 1, the previous *Q* output of the flip-flop is XORed with *Scan_data*, with the result being stored into the flip-flop. Thus, the flip-flop behaves like a *T* flip-flop in this case.

Test vectors and their responses get modified when they are shifted through the modified scan chain architecture consisting of a mix of *D* and *T* flip-flops. Due to the presence of *T* flip-flops, the pattern finally obtained in the scan chain is different from the scan-in pattern. Thus, to get the correct test pattern loaded into the scan chain, the scan-in pattern needs to be modified. In a scan-based design, a test pattern is shifted in simultaneously while the output response of the previous test pattern is shifted out. Thus, to modify the pattern to be shifted in, the response of the previous pattern also needs to be considered. If a particular flip-flop in the scan-chain is a *D* flip-flop, the scan-in value is directly loaded into

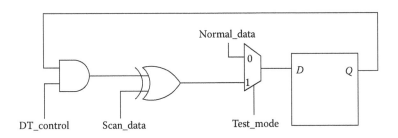

FIGURE 2.2   Scan-cell control between *D*- and *T*-type flip-flops.

it. However, for a *T*-type flip-flop, its content in the current clock cycle is determined by XORing its content with the content of its previous flip-flop in the chain, in the previous clock cycle.

Let O_{ij} represent the content of the *i*th flip-flop at the *j*th cycle of the scan-in process. Then,

$O_{ij} = O_{(i-1)(j-1)}$, if the flip-flop is of type *D*

$= O_{(i-1)(j-1)}$ *XOR* $O_{i(j-1)}$, if the flip-flop is of type *T*.

Now, to apply a test pattern to the circuit, it is necessary that at the end of the shifting process, the scan chain contains that pattern. Let the test pattern to be loaded into an n-cell scan chain be $<v_1, v_2, \ldots v_n>$. Thus, the content of the cells at the *n*th cycle are $O_{1n} = v_1$, $O_{2n} = v_2$, $\ldots O_{nn} = v_n$. To get this pattern, the content of the $(n-1)$th cell at the $(n-1)$th cycle should be

$O_{(n-1)(n-1)} = O_{nn}$, if the *n*th flip-flop is of type *D*

$= O_{nn}$ *XOR* $O_{n(n-1)}$, if the *n*th flip-flop is of type *T*.

In general, for the *i*th cell $(1 \leq i \leq n-1)$, content at the $(j-1)$th cycle $(2 \leq j \leq n)$ should be

$O_{i(j-1)} = O_{(i+1)j}$, if the flip-flop $(i+1)$ is of type *D*

$= O_{(i+1)j}$ *XOR* $O_{(i+1)(j-1)}$, if the flip-flop $(i+1)$ is of type *T*.

For example, suppose that the final pattern to be loaded into a four-cell scan chain is "1101," and the initial content of the scan chain (that is, the response of the previous test pattern) is "0000". Contents of different cells at various cycles have been shown in Table 2.4 corresponding to "DDDD" and "DDTD" chain configurations. It may be noted that, in the all-D chain, the test pattern to be shifted is "1101," while for the "DDTD" configuration, the test pattern to be shifted gets modified to "1111." For the all-D case, there are

TABLE 2.4   Test-Vector Modification for Scan Configurations

Scan Cycle	DDDD Configuration				DDTD Configuration			
0	0	0	0	0	0	0	0	0
1	1	0	0	0	1	0	0	0
2	0	1	0	0	1	1	0	0
3	1	0	1	0	1	1	1	0
4	1	1	0	1	1	1	0	1

TABLE 2.5   Peak-Temperature Results for Scan-Cell Optimization Method

Circuit	Peak Temperature in Degrees Kelvin for	
	All-D Chain	Selective D-T Chain
s5378	318.09	317.01
s9234	330.22	326.76
s13207	329.44	325.11
s15850	325.92	321.96
s38417	320.97	318.35
s38584	321.42	318.43

nine flips in the values of the flip-flops during the scan-in process, whereas, for the "DDTD" case, the number of flips reduces to five.

To design a simple approach to decide upon the types of individual scan flip-flops, consider the don't care filled test set described in Section 2.3. It assumes all flip-flops to be of type D. Now, each of the flip-flops can be modified to T, one at a time. After modifying flip-flop i to type T, the test vector and response pattern through the scan chain are recalculated. The criticality values are evaluated. In the case of getting a better maximum criticality, the ith flip-flop is retained as type T, otherwise it is converted back to D. Table 2.5 shows the corresponding peak temperature results for the approach.

2.5 BUILT-IN SELF TEST

Built-In Self Test (BIST) is a test methodology in which the test pattern generator and response compactor are integrated with the circuit under test. Typically, a *linear feedback shift register* (LFSR) is used as the pseudo-random pattern generator, while a *multiple input signature register* (MISR) is used as the response compactor. After the completion of the test session, the signature obtained in the MISR is shifted out and compared with the stored response. However, to obtain reasonably high fault coverage with pseudo-random patterns, long test sequences often are required. This leads to higher power consumption of the circuit and, consequently, an increase in the peak temperature.

To ensure reduced power consumption and associated temperature rise, a BIST scheme with a modified *test-pattern generator* (TPG) is shown in Figure 2.3. The TPG consists of an LFSR, a k-input AND gate and a T flip-flop. The structure is commonly known as *low transition-random TPG* (LT-RTPG) [3].

Each of k-inputs of the AND-gate is connected to the output of a particular flip-flop in the LFSR. Thus, the tapped LFSR stage outputs are ANDed and the result is fed as input to the T flip-flop. From the behavior of the T flip-flop, the flip-flop output remains unaltered as long as $T = 0$; it toggles when $T = 1$. When the LFSR generates a long random test sequence, the signal probability of any stage of the LFSR is 1/2. Thus, the probability that the

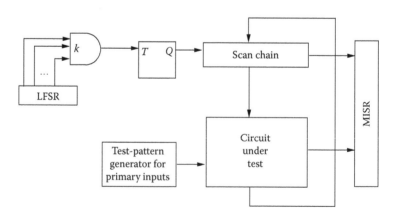

FIGURE 2.3 LT-RTPG architecture.

T flip-flop toggles at any arbitrary clock cycle is $1/2^k$. As a result, a long run of 1 or 0 may be generated at the T flip-flop output with the selection of a large k. The T flip-flop output feeds the scan chain. As the adjacent scan cells are likely to have the same values now, transitions occurring during scan shift get reduced, leading to reduced power consumption.

In the LT-RTPG structure, there can be three parameters that can control the patterns being fed into the scan chain and the associated power and temperature values: primitive polynomial of the LFSR, initial seed value of the LFSR, and the selection of k tapped stages. In the following, a PSO-based scheme is presented to identify good values for these parameters, leading to peak temperature minimization.

2.5.1 PSO-based Low Temperature LT-RTPG Design

Let the LFSR used for pattern generation be of size m. Thus, it can realize a degree m primitive polynomial. In PSO, a solution is represented by a particle. For LT-RTPG design, a particle consists of three parts:

1. *Polynomial*: An array of m bits. The ith bit being "1" indicates that the ith flip-flop output participates in the feedback network of the LFSR. Otherwise, the bit is "0."

2. *Seed*: An array of m bits identifying the initial seed of the LFSR.

3. *Tappings*: An array of m bits. The ith bit indicates whether the ith flip-flop drives the AND gate or not. If the bit is "1," the corresponding flip-flop output forms one of the inputs to the AND gate.

It may be noted that the polynomial has to be a primitive one and that the *Tappings* part of the particle cannot have more than k bits as "1" for a k-input AND gate. Suitable checks are introduced for ensuring these constraints.

TABLE 2.6   Peak Temperature and Fault-Coverage Results for LT-RTPG

Circuit	Normal BIST		LT-RTPG	
	Peak Temp. (°K)	Fault Coverage	Peak Temp. (°K)	Fault Coverage
S5378	432.14	97.32	360.30	96.41
S9234	458.18	87.56	363.15	87.32
S13207	476.89	91.78	368.36	89.91
S15850	477.37	93.41	369.23	92.78
S38417	482.97	88.38	400.43	88.24
S38584	518.96	94.72	408.54	93.43
B14	461.03	91.23	330.40	90.78
B15	455.67	97.67	326.38	97.12
B20	480.73	95.49	335.26	94.26
B21	484.89	92.21	350.49	91.37
Average improvement in LT-RTPG over normal BIST			23.62%	−0.87%

The fitness measure of a particle can be obtained based upon the criticality measure defined in Section 2.2.1. Lower the value of the maximum criticality of circuit blocks, better fit is the particle.

Table 2.6 presents a comparison between normal BIST and LT-RTPG BIST with respect to peak temperature and fault coverage for a number of benchmarks. For all circuits, the AND gate of LT-RTPG has been restricted to have two inputs ($k = 2$). The LFSR is taken to be 10 bits wide. Total number of patterns considered is equal to 10,000. It may be noted that with 10,000 patterns, normal BIST results in higher peak temperature than LT-RTPG; however, fault coverage is a bit poor for LT-RTPG. To reach the same level of fault coverage, more patterns need to be applied to the circuit, which may result in an increase in the peak temperature by some small amount.

2.6 SUMMARY

In this chapter, several circuit-level techniques have been discussed for reducing temperature during testing. Test-pattern reordering has been enumerated using greedy and PSO approaches. ATPG tools leave many of the bits as unspecified in the test vectors. Those have been exploited for temperature reduction via proper filling. Scan-cell optimization approaches have been discussed. Finally, temperature reduction techniques for a BIST environment have been detailed.

REFERENCES

1. A. Dutta, S. Kundu, S. Chattopadhyay, "Test Vector Reordering to Reduce Peak Temperature During Testing", *IEEE India Council International Conference*, 2013.
2. A. Dutta, S. Kundu, S. Chattopadhyay, "Thermal-Aware Don't Care Filling to Reduce Peak Temperature and Thermal Variance During Testing", *Asian Test Symposium*, 2013, pp. 25–30.
3. A. Dutta, S. Chattopadhyay, "Particle Swarm Optimization for Low Temperature BIST", *International Symposium on VLSI Design and Test*, 2015, pp. 1–6.

CHAPTER 3

Test-Data
Compression

3.1 INTRODUCTION

Complexity of digital circuits, in terms of their intended functionality and number of logic elements, has increased significantly over the decades. This has affected the system test time and test cost to a good extent. In order to test for a higher number of possible faults in the circuit, it is necessary to apply more test patterns. These test patterns and their corresponding fault-free responses need to be stored in the *automatic test equipment* (ATE), external to the chip. Testing now encompasses transferring these patterns to the chip, applying them to the *circuit under test* (CUT), getting the responses, and comparing these responses with the fault-free ones. ATE needs to be equipped with large physical memory to accommodate all patterns and responses. However, ATE cost increases significantly with increasing storage capacity. The second important stumbling block is the maximum frequency of the ATE channels engaged in test pattern transportation. Again, the cost of ATE increases with an increase in the frequency of operation. *At-speed* testing demands chips be tested at their desired frequency of operation in the functional mode to catch delay faults. Thus, an ATE operating at lower frequency cannot be used to directly test the CUT. Increasing test data also affect the overall test time, power consumption, and associated heating. *Built-in self-test* (BIST) happens to be an alternative to address these problems. However, BIST may generate a large number of non-detecting, intermediary patterns to reach a significant level of fault coverage.

The more effective solution to the above problem is to make use of data compression policies. Test data compression mechanisms store huge test data (T_D) in a compressed form (T_E) in the ATE memory. Compressed test patterns are transported to the chip through ATE channels. An on-chip decoder is utilized to get back the original test patterns from their compressed versions. Thus, the requirement for ATE memory space is reduced. More test data can be transported in less time, in a compressed format. This leads to test-time reduction. Availability of more test data may enable at-speed testing of the chip. As noted in Chapter 2, test patterns generated by ATPG tools generally contain a large number of don't cares (often more than 95%). These don't care bits are exploited by the compression algorithms to increase similarities between the patterns. However, circuit-level power- and thermal-aware test strategies also utilize these don't cares (as noted in Chapter 2). Thus, there is a trade-off between the degree of compression and temperature distribution of the CUT that can be achieved via different don't care bit filling policies. This chapter first shows the existence of this variation in peak temperature of a circuit corresponding to different don't care bit filling techniques targeted toward test-data compression. This has been followed by the enumeration of a strategy that can perform an effective tradeoff. This has been augmented by another technique that can reduce peak temperature further without compromising on compression rate. All these strategies have been developed around a static dictionary-based test data compression policy.

In Section 3.2, a brief overview of such dictionary-based compression policy is presented. Section 3.3 enumerates a technique for dictionary construction using clique partitioning. Section 3.4 presents a trade-off between peak temperature and compression. Section 3.5 discusses reducing peak temperature further without compromising on compression ratio.

3.2 DICTIONARY-BASED TEST-DATA COMPRESSION

A typical static dictionary-based test compression policy is shown in Figure 3.1. Here, test pattern bits are shifted serially to the chip from the ATE. The bit pattern is used to index into a dictionary forming a part of the on-chip decoder. Actual test bits are retrieved from the dictionary and fed into the scan chains of the CUT.

Let there be n test patterns, each of length L. Thus, the total number of test bits is equal to $T_D = n \times L$. Also assume that there are m scan chains in the CUT. In that case, m bits of test data are to be transferred to the CUT simultaneously. If the length of each scan chain is l, the number of test slices per pattern is also l. Thus, $L = l \times m$. For example, consider a test pattern "101101011011" of length $L = 12$. If the CUT has three scan chains ($m = 3$), each of length 4 ($l = 4$), the part of the pattern in the first chain is "1011," second chain is "0101," and the third chain is "1011." The least significant test slice is formed by considering the least significant bits of the three scan chain contents, that is, "111," while the successive slices are "101," "010," and "101," respectively.

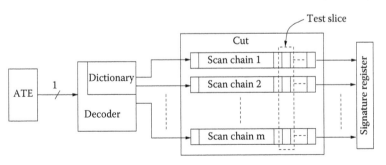

FIGURE 3.1   Dictionary-based test-data compression architecture.

Here, it may be noted that the slice "101" occurs twice. Thus, in the dictionary, it can have a single entry, reducing the test data size. Such slices are called compatible slices.

In general, two slices A and B, each of size m, are said to be compatible if, for all i ($1 \le i \le m$), $A_i = B_i$ or at least one of A_i and B_i is a don't care. For a set of such compatible slices, only one representative slice may be stored in the dictionary. Due to the size constraint of the dictionary, only some such representative slices can be stored there. In particular, if the dictionary can accommodate D slices in it, D largest compatible sets of slices can be stored there. For such slices, the bit sequence transferred from the ATE is an index into the dictionary. Other slices are stored in uncompressed form in the ATE memory directly. Whenever a slice is transferred from ATE to the decoder, the first bit (known as the prefix bit) identifies its type. A prefix bit of "0" signifies a dictionary slice. In that case, the next few bits constitute the index to the dictionary. The decoder concatenates those bits to form the index, accesses the corresponding dictionary location, and its content is fed as the next slice to the scan chains. On the other hand, a prefix bit "1" indicates an uncompressed slice. The next few bits transferred from the ATE are the slice bits. These bits are directly sent to the corresponding scan chains. Thus, the slices for which representatives are available in the dictionary are denoted by a code of length $\lceil \log_2 |D| \rceil + 1$, while for an uncompressed slice, the code length becomes $m + 1$. It may be noted that the choice of D largest compatible sets of slices is the most crucial issue in determining the achievable compression ratio.

3.3 DICTIONARY CONSTRUCTION USING CLIQUE PARTITIONING

To determine the slice entries to be put onto the dictionary, a clique partitioning-based formulation is often used [1]. For this, an undirected graph $G = (V, E)$ is constructed. Each test slice corresponds to a vertex $v \in V$. Two vertices v_i and v_j are connected by an edge $e_{ij} = (v_i, v_j)$ if their corresponding slices are compatible.

For example, the slice "1X0X10" is compatible with the slice "100X10." Hence, the corresponding vertices will have an edge in E. These two slices have a common slice "100X10" that can represent both. A clique of a graph is a completely connected component of it such that, between any pair of vertices in the component, there is an edge. The graph G can be partitioned into a number of disjoint cliques. Corresponding to each clique, a representative slice can be put into the dictionary. Due to size limitations of the dictionary, it is not possible to accommodate representatives for all of the cliques. If the dictionary size is D, the largest D cliques can be put into it. Thus, the clique partitioning process should try to identify cliques with large numbers of nodes in them. However, the clique partitioning problem to identify the smallest number of cliques partitioning the graph is *NP-Hard*. Several heuristics have been proposed in the literature to address the problem. The strategies typically start with the vertex having the largest number of incident edges (that is, the highest degree). The corresponding clique is constructed. Next, all vertices and edges of this clique are removed from G, and the process is repeated by identifying the highest degree surviving vertex. However, the way the start vertex is selected for clique formation may have a significant effect on the cliques produced. The alternatives can be explored by slightly modifying the heuristic mentioned earlier. Instead of starting the clique formation process with a node of highest degree, at the first place, each vertex of G can be considered as the start node. Thus, for a graph with n nodes, it is expected to generate n different ways of partitioning the graph. Generated cliques may not always be unique, but for a large number of cases, it is expected to be so. The overall algorithm to perform clique partitioning starting at vertex *start* is presented next.

Algorithm Clique_Partition

Input: Graph G, Start vertex *start*.
Output: The cliques.

Begin

Step 1: Copy G into a temporary data structure G'.

Step 2: Establish a subgraph consisting of all vertices connected to start. Copy this subgraph to G' and add *start* to a set C. G' does not include *start*.

Step 3: If G' is not empty, set *start* to the node with the maximum degree in G'; Go to Step 2. Otherwise a clique C has been formed.

Step 4: Remove the vertices in clique C from G and copy $G - C$ to G'. Set start to the vertex with the maximum degree in G'. Go to Step 2 and repeat until G becomes empty.

End.

EXAMPLE 3.1

Consider a set of eight test slices v_1, v_2, ... v_8, shown in Figure 3.2a, each with a width of 8 bits. The corresponding graph has been shown in Figure 3.2b. Taking v_1 as the start vertex, the clique set becomes $\{(v_1, v_2, v_3, v_4), (v_5), (v_6), (v_7), (v_8)\}$.

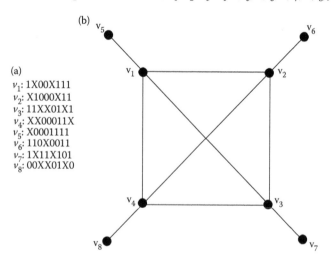

(a)
v_1: 1X00X111
v_2: X1000X11
v_3: 11XX01X1
v_4: XX00011X
v_5: X0001111
v_6: 110X0011
v_7: 1X11X101
v_8: 00XX01X0

FIGURE 3.2 (a) An example test set and its graph, (b) graph corresponding to the test set.

TABLE 3.1   Variation in Peak Temperature and Compression Ratio Among
Clique Sets

Circuit	Test Bits (TD)	Clique Sets	Peak Temperature (°K)		Compression Ratio	
			Maximum	Minimum	Maximum	Minimum
s13207	165200	5192	352.1	343.2	73.4	73.2
s15850	76986	2520	349.8	338.7	67.9	67.0
s38417	164736	5148	351.2	342.5	45.5	43.7
b14	211074	6858	365.2	357.2	56.8	55.0
b15	222102	7312	358.7	334.7	73.3	73.0

Whereas, with v_5 as the start vertex, the clique set becomes $\{(v_1, v_5), (v_2, v_3, v_4), (v_6), (v_7), (v_8)\}$. For the first set of cliques, the representative patterns are {11000111, X0001111, 110X0011, 1X11X101, 00XX01X0}. As noted in Chapter 2, each pattern may result in a different thermal behavior of the chip.

Table 3.1 shows the variation in peak temperature and compression ratio that could be observed in the partitions generated for a set of benchmark circuits. Each circuit has been assumed to have 32 scan chains. Corresponding to any circuit, the number of distinct clique sets is noted. For each of these clique sets, the compression ratio has been computed. The thermal behavior observed by applying the resulting test pattern sets is noted. The maximum and the minimum temperature values among all the cliques are reported in the table. For example, for the circuit s38417, peak temperature shows a difference of 7.7 degrees between the maximum and the minimum values, while compression ratio has a variation of 1.8%.

3.4 PEAK TEMPERATURE AND COMPRESSION TRADE-OFF

The strategy enumerated in the last section suffers from two major drawbacks. The first one is that the exploration of clique sets relies on compression-aware don't care filling only. Two slices are merged by utilizing these don't care bits. It does not look into the thermal effects of the resulting patterns. In Chapter 2 [2], thermal-aware don't care bit filling techniques were discussed which also attempt to utilize these don't cares. Secondly, thermal simulation of all the huge number of clique sets (as noted in Table 3.1) is a time-consuming operation and thus not practical for reasonably complex circuits. With this background, a strategy is reported next that can control the amount of thermal-aware and compression-aware patterns to be mixed in the final test set [3].

As noted earlier, test pattern set generated by an ATPG tool contains a large number of don't care bits. Let this pattern set be *TP*. Using the thermal-aware don't care bit filling technique noted in Chapter 2, the don't cares are filled to generate test set TP_{TH} from *TP*. Depending upon the scan chain configuration, test slices are generated from TP_{TH}. Let these thermal-aware slices constitute the set SI_{TH}. The original test set *TP* is now processed by the clique partitioning strategy noted in Section 3.3. This generates the compression-aware test slices SI_{CM}. Compression-aware slices are sorted in the descending order of their corresponding clique sizes. To get a mix of thermal- and compression-aware slices in the final test set, a weight factor *Wt* ($0 \leq Wt \leq 1$) is considered. The weight factor determines the percentages of slices coming from SI_{TH} and SI_{CM} to form the final test set. If *n* is the number of test patterns in *TP* and *l* is the number of slices per pattern, the total number of test slices in *TP* is equal to $n \times l$. Out of these, $TSC = Wt \times n \times l$ number of slices will be made compression-aware. To construct the final test set, the process starts with TP_{TH}, with the corresponding slice set SI_{TH}. From the set SI_{TH}, *TSC* slices are replaced by taking the topmost *TSC* slices from the sorted compression-aware test slice set SI_{CM}. This modification leads to a new set of test slices SI_{NEW} and the corresponding test pattern set TP_{NEW}. A clique partitioning heuristic is now run on this test set to generate new cliques, out of which the topmost $|D|$ entries can be accommodated in the dictionary of size $|D|$. Peak temperature and the compression ratio corresponding to this test set are computed. The whole procedure is noted as an algorithm next. The weight factor *Wt* determines the degree of compression-aware slices mixed into the thermal-aware test set. Setting $Wt = 1$ will make the set completely compression-aware, ignoring the peak temperature. Setting $Wt = 0$ prohibits any thermal-aware test slice from getting replaced by compression-aware ones.

Algorithm Temperature_Compression_Trade-Off

Input: Test pattern set TP, Weight factor Wt.
Output: Mixed test pattern set TP_{NEW}.
Begin
Step 1: Compute thermal-aware test set TP_{TH} from TP using thermal-aware don't care filling (Section 2.3).
Step 2: Compute test slices SI_{TH} corresponding to TP_{TH}.
Step 3: Perform clique partitioning on input test set TP.
Step 4: Compute compression-aware test slices SI_{CM} of TP.
Step 5: Sort slices of SI_{CM} on descending order of corresponding clique sizes.
Step 6: $TSC = Wt \times n \times l$.
Step 7: Replace TSC slices from SI_{TH} by the topmost TSC slices from SI_{CM}. Let the new slice set be SI_{NEW}. Compute corresponding test set TP_{NEW}.
Step 8: Perform clique partitioning of TP_{NEW} to get $|D|$ dictionary entries. Modify TP_{NEW} accordingly.
Step 9: Compute the compression ratio of TP_{NEW} with respect to TP. Compute the peak temperature for TP_{NEW}.
Step 10: Output TP_{NEW}, compression ratio and peak temperature.
End.

Table 3.2 notes the results of the temperature-compression trade-off algorithm on a set of benchmarks. The dictionary size has been assumed to be $|D| = 128$. Two different scan chain lengths, 64 and 128, have been considered. The weight factor Wt has been varied from 0 to 1. For $Wt = 0$, all don't cares are filled, aiming for temperature reduction. As can be observed from the table, this results in the minimum temperature values; however the corresponding compression ratios are very poor. For $Wt = 1$, compression ratios

TABLE 3.2   Peak Temperature and Compression Ratio (CR) Tradeoff

Circuit	Scan Chains		Weight Factor Wt				
			0.0	0.2	0.5	0.8	1.0
s38584	64	Temp. (°K)	347.85	363.61	378.05	385.36	386.91
		CR (%)	18.82	25.06	46.35	67.37	70.13
	128	Temp. (°K)	376.87	395.19	410.09	418.44	418.77
		CR (%)	9.39	23.55	49.68	67.97	70.87
s38417	64	Temp. (°K)	352.05	368.22	374.36	377.58	378.46
		CR (%)	10.89	20.79	37.19	41.73	43.30
	128	Temp. (°K)	390.49	406.06	417.52	419.60	420.88
		CR (%)	9.28	24.33	33.58	36.57	36.57
s13207	64	Temp. (°K)	338.70	362.27	379.18	391.45	394.73
		CR (%)	63.28	63.83	66.95	75.80	85.01
	128	Temp. (°K)	352.33	385.72	417.02	432.72	436.18
		CR (%)	63.50	66.38	72.05	83.40	91.30
s15850	64	Temp. (°K)	341.93	359.48	372.71	379.61	379.32
		CR (%)	38.79	41.76	50.95	73.43	75.84
	128	Temp. (°K)	365.53	381.57	390.85	396.35	394.37
		CR (%)	35.25	43.94	64.64	80.84	81.59
b14	64	Temp. (°K)	350.67	357.78	375.98	387.97	389.43
		CR (%)	26.80	29.01	44.32	44.74	43.78
	128	Temp. (°K)	379.70	380.37	401.17	416.69	416.96
		CR (%)	38.10	39.40	45.57	55.76	61.72
b15	64	Temp. (°K)	332.41	352.09	370.87	381.08	384.21
		CR (%)	33.98	41.60	54.58	72.12	80.65
	128	Temp. (°K)	349.24	379.31	398.11	408.85	409.99
		CR (%)	18.71	31.59	52.63	73.53	73.53

becomes impressive; however, temperature increases significantly. Other values of Wt produce results balancing the optimal results for peak temperature and compression ratio. The results, in general, bring out the trade-off between the two. However, in some cases, a trade-off is missing. For example, for circuit b14, for Wt larger than 0.8, the compression ratio has decreased for the 64 scan chains case. This can be attributed to the heuristic nature of the trade-off algorithm.

From Table 3.2, it can also be observed that a dictionary of size $|D| = 128$ often provides better compression than a smaller sized

TABLE 3.3   Dictionary Area Overheads

| Circuit | Circuit Area (μm²) | Scan Chains | Dictionary Area Overheads (μm²) | | | |
			$D = 128$	In %	$D = 64$	In %
s38584	128622.6	64	3942.00	3.06	2050	1.60
		128	5835.00	4.54	2982	2.32
s38417	133961.1	64	3767.00	2.81	1886.33	1.41
		128	4367.33	3.26	2068.66	1.54
b14	29902.81	64	3890.00	13.01	2008.00	6.71
		128	4961.00	16.59	2942.66	8.33
b15	56876.05	64	3747.00	6.59	1995.21	3.51
		128	4895.40	8.61	2354.96	4.14

dictionary, such as $|D| = 64$. However, a larger-sized dictionary has higher overhead in terms area. Table 3.3 shows area overheads for dictionaries of different sizes for circuits having different numbers of scan chains. Area values have been obtained by synthesizing both circuits and dictionaries using a Synopsys Design Vision compiler with a Faraday 90 nm technology library. It may be observed that the area overheads are quite small, suggesting the usage of reasonably large dictionaries.

3.5 TEMPERATURE REDUCTION WITHOUT SACRIFICING COMPRESSION

In the last section, a trade-off has been performed between thermal- and compression-aware test slices. This has the potential to explore a range of don't care filled test sets with various peak temperature and compression ratio values. However, there can be further possible reductions in peak temperature of a test set without affecting the associated compression ratio [4]. This may be important from the following aspects:

1. Sacrificing compression ratio increases test set size, which eventually increases the overall test time. This may affect the time-to-market for the chip. Hence, reduction in compression ratio may not always be desirable.

2. Compression-aware test slices may still have don't care bits present in them. This may happen if all slices in a clique have at least one don't care bit at same position. The representative slice, in that case, will also have don't cares. Such don't cares may be filled in a thermal-aware manner.

3. If the dictionary size is small, generating a large number of compression-aware slices is of no use. More specifically, in the Algorithm Temperature_Compression_Trade-Off, if the TSC value is larger than the dictionary size $|D|$, the extra slice replacements do not aid in improving the compression ratio further. They may better be continued as thermal-aware slices for peak temperature reduction.

The aforementioned concerns lead to an improved strategy of test set formulation. Given the compression-aware clique list CL_{CM}, the corresponding slice set SI_{CM}, thermal-aware test slices SI_{TH}, and the dictionary size D, a final test set is generated that addresses the shortcomings mentioned earlier. Because the first and foremost objective is not to compromise on the compression ratio, the representative slices corresponding to the $|D|$ largest cliques of CL_{CM} are taken into SI_{FINAL}. To ensure the generation of the complete test set, some more slices are necessary, because all test patterns need not be covered by these $|D|$ slices. To complete the set, representative slices are chosen in a thermal-aware fashion from the set SI_{TH}. The slices selected from SI_{TH} are completely filled up. However, the $|D|$ slices in the dictionary may still contain some don't cares. Setting them to either "0" or "1" does not affect the compression ratio. If one such bit in a representative slice is set to "0" ("1"), in the final test set, all of the slices represented by it will have "0" ("1") in that position. This may affect the thermal behavior of the test patterns containing that slice. Moreover, from the thermal viewpoint, a don't care bit of a slice may be filled up differently in different patterns. Thus, while making the decision about a bit in a test slice of a dictionary, the corresponding slices represented by it need to be considered. If, for a particular don't care bit, the thermal-aware slices of the represented test slices suggest a "0" value in more cases than a "1" value, the bit should be filled with a "0"; otherwise, it should be filled with a "1." The overall procedure has been noted as an algorithm next.

Algorithm Temperature_Optimization

Input: Test pattern set TP, Compression-aware clique list CL_{CM}, Compression-aware slices SI_{CM}, Thermal-aware slices SI_{TH}, Dictionary size $|D|$.

Output: Final test pattern set TP_{FINAL}.

Begin

Step 1: Select $|D|$ largest cliques from CL_{CM}.

Step 2: Create dictionary with representative test slices corresponding to the selected $|D|$ cliques.

Step 3: Copy representative slices from SI_{CM} for these $|D|$ cliques into SI_{FINAL}.

Step 4: Select slices from SI_{TH} to be put into SI_{FINAL} to ensure generation of complete test set TP.

Step 5: For all $|D|$ dictionary entries do

>**begin**
>
>>For $j = 1$ to m do   // m = slice size = number of scan chains
>>
>>**begin**
>>
>>>If jth bit is a don't care then
>>>
>>>**begin**
>>>
>>>>Identify represented slices for this representative slice.
>>>>
>>>>$Count\_zero = Count\_one = 1$.
>>>>
>>>>For each represented slice s do
>>>>
>>>>**begin**
>>>>
>>>>>Let t be the thermal-aware slice corresponding to s.

If bit $t[j] =$ "0" then

Count_zero $=$ Count_zero $+ 1.$

Else

Count_one $=$ Count_one $+ 1.$

end;

If Count_zero $>$ Count_one then

Fill the jth bit of all represented slices with "0" in $SI_{FINAL}.$

Else

Fill the jth bit of all represented slices with "1"in $SI_{FINAL}.$

end;

end;

end;

Step 6: Form the final test pattern set TP_{FINAL} from $SI_{FINAL}.$
End.

The next example illustrates the operation of the algorithm.

EXAMPLE 3.2

Consider a test set with eighteen test slices. When full test compression is targeted, let those be merged into nine representative slices as follows.

No.	Representative Slice								Slice Indices Merged
1	1	0	0	1	0	0	0	1	1, 2
2	1	0	X	0	0	0	X	1	3, 7
3	1	0	1	1	0	0	1	0	4, 5, 6, 8
4	0	1	X	X	0	0	0	X	9
5	X	1	0	X	1	1	X	X	10
6	0	0	1	0	0	1	X	0	11, 12, 13, 14
7	0	1	0	1	X	0	1	X	15, 16
8	0	1	0	X	X	X	1	X	17
9	X	1	X	1	X	0	1	0	18

The corresponding eighteen thermal-aware slices are as follows.

Slice Index	Thermal Slice							
1	1	0	1	1	0	1	0	1
2	1	0	0	0	0	0	0	1
3	1	0	0	0	0	0	1	1
4	1	0	1	1	0	0	0	0
5	1	0	1	1	0	0	0	0
6	1	0	0	0	0	0	1	0
7	1	0	0	0	0	1	1	1
8	1	0	1	1	0	0	1	0
9	0	1	0	0	0	0	0	0
10	1	1	0	1	1	1	1	1
11	1	0	0	0	0	1	0	0
12	0	0	0	0	0	1	0	0
13	0	0	1	0	0	1	0	0
14	0	0	1	0	0	1	1	1
15	0	1	0	1	1	0	1	1
16	1	1	1	1	1	0	1	1
17	0	1	0	1	1	1	1	1
18	1	1	0	1	0	0	1	0

Assuming that the dictionary is of size four, the representative slices corresponding to the four largest cliques are put into it. Thus, the content of the dictionary is

$$1 \quad 0 \quad 1 \quad 1 \quad 0 \quad 0 \quad 1 \quad 0$$
$$0 \quad 0 \quad 1 \quad 0 \quad 0 \quad 1 \quad X \quad 0$$
$$1 \quad 0 \quad 0 \quad 1 \quad 0 \quad 0 \quad 0 \quad 1$$
$$1 \quad 0 \quad X \quad 0 \quad 0 \quad 0 \quad X \quad 1$$

Consider the X-bit in the second dictionary element. The slice represents the test slices 11, 12, 13, and 14 in the dictionary. The thermal-aware versions of the slices 11, 12, and 13 contain a "0" at that particular bit position, while for slice 14, the bit is "1." Thus, in the *Algorithm Temperature_Optimization, Count_zero* becomes more than *Count_one*. Hence, in the dictionary, the bit is set to "0." For the last slice in the dictionary, the slices represented by it are 3 and 7. In the thermal-aware version of these slices, the bit at the position of first "X" (from LSB) is "1," while the second "X" is "0." Thus, in the dictionary, the slice becomes "10000011." Hence, the final eighteen slices (SI_{FINAL}) produced by the algorithm are as follows.

Slice Index	Slice							
1	1	0	0	1	0	0	0	1
2	1	0	0	1	0	0	0	1
3	1	0	0	0	0	0	1	1
4	1	0	1	1	0	0	1	0
5	1	0	1	1	0	0	1	0
6	1	0	1	1	0	0	1	0
7	1	0	0	0	0	0	1	1
8	1	0	1	1	0	0	1	0
9	0	1	0	0	0	0	0	0
10	1	1	0	1	1	1	1	1
11	0	0	1	0	0	1	0	0
12	0	0	1	0	0	1	0	0
13	0	0	1	0	0	1	0	0
14	0	0	1	0	0	1	0	0
15	0	1	0	1	1	0	1	1
16	1	1	1	1	1	0	1	1
17	0	1	0	1	1	1	1	1
18	1	1	0	1	0	0	1	0

To get an idea about the range of temperature reductions that could be achieved without sacrificing the compression ratio, the algorithm has been applied on benchmark circuits, as noted in Table 3.4. Dictionary size has been taken to be 128. For the sake of comparison, temperature values correspond to $Wt = 1$ in Algorithm Temperature_Compression_Trade-Off. These have been compared with the temperature values obtained by the Temperature_Optimization algorithm. Percentage reductions in peak temperature values are noted. Table 3.4 shows that up to a

TABLE 3.4   Temperature Optimization Without Sacrificing Compression Ratio

Circuit	Scan Chains	Temperature_Compression_ Trade-Off (Wt = 1) (K)	Temperature_ Optimization (K)	Percentage Reduction
s38584	64	386.91	376.76	25.99
	128	418.77	411.11	18.28
s38417	64	378.46	373.30	19.54
	128	420.88	412.10	28.39
b14	64	389.43	375.90	34.91
	128	416.96	406.10	29.15
b15	64	384.21	368.14	31.02
	128	409.99	394.24	25.93

34% reduction in peak temperature value can be obtained without sacrificing the compression ratio.

3.6 SUMMARY

This chapter has presented techniques to utilize thermal-aware procedures while constructing static dictionaries for test compression. Clique partitioning approaches normally used to identify compatible test slices and perform their merging for dictionary construction have been modified to integrate thermal-aware test slices with them. Trade-off has been performed while filling the don't care bits in test patterns to give relative importance to compression ratio and peak temperature optimization. A scheme has also been shown to achieve temperature improvements without sacrificing the compression ratio. Experimental results with respect to benchmarks have established the suitability of the schemes.

REFERENCES

1. L. Li, K. Chakrabarty, N.A. Touba, "Test Data Compression using Dictionaries with Selective Entries and Fixed-Length Indices", *ACM Transactions on Design Automation of Electronic Systems*, Vol. 8, No. 4, 2003, pp. 470–490.
2. A. Dutta, S. Kundu, S. Chattopadhyay, "Thermal-Aware Don't Care Filling to Reduce Peak Temperature and Thermal Variance During Testing", *22nd Asian Test Symposium*, 2013, pp. 25–30.

3. R. Karmakar, S. Chattopadhyay, "Thermal-Aware Test Data Compression using Dictionary Based Coding", *28th International Conference on VLSI Design*, 2015, pp. 53–58.
4. R. Karmakar, S. Chattopadhyay, "Temperature and Data-Size Tradeoff in Dictionary Based Test Data Compression", *Integration, the VLSI Journal*, Vol. 57, 2017, pp. 20–33.

CHAPTER 4

System-on-Chip Testing

4.1 INTRODUCTION

System-on-chip (SoC) based design is a paradigm shift in the VLSI design process. In SoC, third-party *intellectual property* (IP) cores are integrated into the chip by a system integrator. Previously designed modules are reused to achieve the system functionality. This has greatly reduced design time, as the modules are not needed to be designed from scratch. However, for the test engineers, challenges increase manyfold. Due to the inherent impurity of silicon wafers, each and every core manufactured on the die needs to be tested exhaustively (even if the manufacturing process is 100% accurate) using the test-pattern sets provided by the IP vendors. This poses a difficulty as the input-output points of individual cores are often not directly accessible from the pins of the chip. To circumvent this problem, on-chip test resources, also known as *test-access mechanism* (TAM) wires, are introduced. Cores are distributed to sets of TAM wires to transport-test patterns and responses. The test engineer is limited by the following:

1. Limited number of channels of the *automatic test equipment* (ATE) to transport-test patterns to the chip.

2. Limited number of on-chip TAM wires.

3. Different number of input-output pins for different cores. Scan chain lengths also vary in nature.

4. Huge amount of test data to be stored in the ATE and transported to/from the chip.

5. High-test power requirement due to parallel testing of power-hungry cores, which may not be active simultaneously in the functional mode of the chip. Thus, design-level power budgets may not work in the test mode.

6. Chip floorplan, power density of cores, core sizes, and their relative position play important roles to determine the temperature of the chip. A test schedule without taking care of thermal issues may lead to violation of the permitted peak temperature of the chip.

The rest of this chapter is organized as follows. Section 4.2 introduces the SoC test problem. A superposition principle-based thermal model is presented in Section 4.3. Section 4.4 details the test-scheduling strategy. Experimental results are noted in Section 4.5. Section 4.6 summarizes the chapter.

4.2 SOC TEST PROBLEM

Let an SoC design consist of N cores: C_1, C_2, ... , C_N. Core C_i ($1 \leq i \leq N$) has the following attributes related to testing:

- PI_i: primary input terminals

- PO_i: primary output terminals

- BI_i: bidirectional (input and output) terminals

- S_i: scan chains, with length (number of scan flip-flops) of chain k as l_{ik}

- P_i: number of test patterns

On the other hand, a fixed number of TAM wires are incorporated into the SoC. A core is allotted a subset of these TAM wires to transport its test patterns from ATE to the core and the responses from the core to the ATE for analysis. Allocation of TAM wires is exclusive, in the sense that at a point of time, a TAM wire can carry test pattern/response bits of a single core only. Larger cores typically have hundreds of terminals, whereas the total number of TAM channels available depends on the limited number of SoC pins. To reduce the overall test time for the SoC, a number of cores may be tested in parallel. Each such core requires a subset of TAM wires allocated to it to carry the test patterns and responses. If l TAM wires are allocated to a core, test patterns for the core reach it via these lines only. However, in general, the total count of primary inputs, bidirectional lines, and scan chain inputs is much larger than this l. The same is true for the outputs. To match the inputs and outputs with l, a test wrapper [1] is designed. The wrapper groups the primary inputs, primary outputs, internal scan chains, and bidirectional lines into l wrapper scan chains. The wrapper scan chain design is performed in a manner such that their lengths are almost the same. If the scan-in lengths of l wrapper chains are SI_1, SI_2, ... SI_l and the scan-out lengths are SO_1, SO_2, ... SO_l, for a test pattern set with p patterns, the overall test time is given by [1]:

$$T = (1 + \max(SI_i, SO_i)) \times p + \min(SI_i, SO_i), 1 \leq i \leq l \quad (4.1)$$

If the total number of TAM lines in an SoC is W, the value of l can be varied in the range 1 to W, each of them leading to a corresponding test time. However, for a range of values of l, test time values may remain unaltered. Thus, in a pareto-optimal design principle, only the pareto-optimal points (values of l at which the test time changes) need to be considered. While test time varies with l, the power and energy consumption pattern also changes. Shifting of test and response bits through scan chains affects these values. Accordingly, heating of the cores also gets affected. To estimate the temperature variations, a thermal simulator needs to be run with the chip floorplan and test power profiles of cores as inputs. However, this thermal simulation is time consuming. In the following, a superposition principle-based strategy has been elaborated that can be utilized to get thermal effects without repeated explicit thermal simulations.

4.3 SUPERPOSITION PRINCIPLE-BASED THERMAL MODEL

The superposition principle takes advantage of the linearity of thermal models used in thermal simulators. The proposed model uses a thermal simulator for each core individually in different possible conditions and creates corresponding thermal databases. This database information is used to generate a thermal-aware test schedule for the cores. The resulting superposition model, although it does not use online thermal simulation, produces the same temperature profile as via the thermal simulators. The superposition principle includes both heating and cooling of cores, subject to the conditions of its neighbors.

To get the exact thermal behavior of a core, different conditions need to be considered which can change the temperature of a core. The temperature of a core C_i may increase because of the following activities:

1. From the beginning of the test session for the SoC up to the time point at which testing of core C_i starts, C_i consumes leakage power. This results in an increase in the initial temperature at which testing of C_i begins.

2. During the testing of C_i, it consumes dynamic power and thus its temperature increases.

3. If any other core C_j is tested in parallel with C_i, because of lateral spreading of heat from C_j, the temperature of C_i will increase. This heating depends on the distance between the cores in the floorplan.

4. The initial temperature of core C_i, before testing initiates, increases due to the effect of all other cores for which testing has been scheduled before C_i.

Figure 4.1 shows different components controlling the temperature profile of a core (say, core 1) in a SoC during its test session. Figure 4.1a shows the typical increase in core temperature due to leakage power consumption before the core is actually scheduled for testing. As it can be observed from the figure, after some time, the temperature saturates to a value because, from that point onward, heat generated via leakage power consumption balances with the heat transferred to the heat sink. Figure 4.1b shows the extra heating effects on core 1 if its testing is scheduled in parallel with another core (say, core 2). Lateral heat spreading of core 2 causes an extra temperature rise in core 1. An important observation from the figure is that the temperature rises to its peak value quickly, and after that, the temperature profile is almost flat. It is justified from the fact that there are not many fluctuations in the middle of the power profile as the number of switches (in scan flip-flop values) does not vary much from one clock cycle to the next, causing little variations in the dynamic power consumption in the middle. Figure 4.1c shows the effects of already-scheduled cores on the initial temperature of a core considered to be scheduled next. When core 1 is idle and core 2 is being tested, the ambient temperature of core 1 starts increasing because of the lateral heat spreading from core 2. When core 2 has just finished its testing, the ambient temperature of core 1 is at its peak. If testing of core 1 commences immediately, the resulting overall temperature of core 1 may be high enough to violate the temperature budget of the SoC. However, if testing

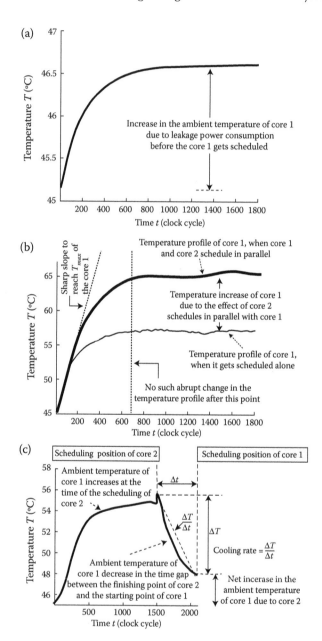

FIGURE 4.1 Effects on temperature profile of core 1 due to: (a) leakage of core 1, (b) parallel testing of cores 1 and 2, and (c) core 2 scheduled earlier than core 1.

of core 1 starts after a certain interval from the completion of testing of core 2, in this time gap, the temperature of core 1 will reduce, possibly leading to a more thermal-safe schedule than in the previous case. Any thermal model should be able to handle all of these conditions.

The objective of the superposition principle-based thermal model is to sum up the individual effects of other cores on the temperature of a particular core to get their cumulative effect on the temperature of the target core. The superposition model works in the following steps:

1. Calculate the increase in the initial temperature of a core due to its own leakage power consumption.

2. Next, for each core C_i ($1 \leq i \leq N$), calculate its temperature rise when it is tested alone. This is followed by the calculation of the extra increase in the temperature of core C_i due to testing of core C_j ($1 \leq j \leq N$, $i \neq j$), when C_i is tested in parallel with C_j. A thermal database is created. It is a matrix of size $N \times N$. For each row i of the matrix, the element (i, i) stores the value of temperature increase of the ith core, when it is tested alone. All other (i, j) elements store the values of extra increases in the temperature of C_i due to parallel testing of C_j with C_i. The superposition principle is applied to calculate the actual increase in the temperature of core C_i considering all other cores tested in parallel with C_i.

3. The next step is to compute the effects of cores tested previously on the core to be scheduled next. For each core C_i ($1 \leq i \leq N$), it is required to calculate its initial temperature increase due to core C_j ($1 \leq j \leq N$, $i \neq j$), scheduled prior to C_i. Let the temperature of C_i when C_j has just finished testing be T_l. The cooling rate of C_i can be computed as the slope of the temperature reduction curve, shown in Figure 4.1c. The part of the curve can be approximated using a linear

approximation. If the decrease in temperature is ΔT over the time period Δt, the cooling rate CR is given by

$$CR = \frac{\Delta T}{\Delta t}.$$

Thus, if testing of C_i commences after time Δt, the initial temperature turns out to be $(T_1 - \Delta T)$. This computation is repeated for each sequence of two cores $<C_i, C_j>$, implying the situation in which C_j is tested after C_i has finished. The corresponding cooling rate is noted in the database. The superposition principle is applied to calculate the actual increase in temperature of C_i, considering all C_j cores scheduled prior to C_i.

Assuming that core C_i is scheduled to start testing at time $StartC_i$ and finish at time $EndC_i$, cores C_i and C_j are considered to be tested parallel if $StartC_i \geq StartC_j$ and $EndC_j \geq StartC_i$.

4.4 TEST-SCHEDULING STRATEGY

The thermal-aware test-scheduling problem can be stated as follows.

Suppose a SoC with N cores $C_1, C_2, \ldots, C_N$ is to be tested with a maximum of W_{max} TAM wires, a maximum power consumption P_{max} and a maximum temperature limit T_{max}. The test-scheduling problem is to allocate TAM wires and test times to the cores such that the total *test application time* (TAT) for the SoC is minimized, while the power consumption during testing remains below P_{max} and the maximum temperature of any core does not cross the upper limit T_{max}.

The overall test-schedule generation process works in two phases. In the first phase, pareto-optimal test points are generated for each core, corresponding power profiles are computed [2], and the thermal databases are created. In the second phase, a *particle swarm optimization* (PSO) based strategy is used to select one test point for each core. A heuristic has been used to generate a schedule for each core, satisfying the power and thermal constraints for the SoC.

4.4.1 Phase I

In this phase, first pareto-optimal test points are generated (as discussed in Section 4.2) using a wrapper design methodology (detailed in [1]). For each TAM width allocation to a core, a test wrapper is designed. It results in a corresponding test time for the core (noted in 1). Thus, for a core, test alternatives can be considered as test rectangles. For each rectangle, the height corresponds to the number of TAM wires allotted and the width is the test time.

Apart from test time, each wrapper design controls the power consumption of the core during test. Scan chain organization varies with the wrapper design. Thus, as the test patterns and responses are shifted through the scan chains, different organizations of these chains lead to different numbers of scan transitions and associated power consumption by the cores. If the dynamic power consumed by a core is DP and the average number of transitions in the total test time of the core is ATR, then for any time interval j, seeing TRN_j scan transitions, power consumed by the core can be approximated as $(TRN_j/ATR) \times DP$. Depending upon the granularity of the thermal simulation process, power profiles of a core can be taken at some time intervals. Considering such a window interval, the peak power value in the interval is taken as the power consumption for the entire window.

Next, four thermal databases, as follows, are constructed.

- *TD1*: *Thermal Database 1* stores the values of increases in the initial temperature due to leakage power consumption of each core.

- *TD2*: *Thermal Database 2* stores the temperature values of self-testing as well as parallel testing of cores.

- *TD3: Thermal Database 3* stores the maximum increase in the initial temperature of the core due to earlier scheduling of another core.

- *TD4: Thermal Database 4* stores the initial temperature of a core after another core completes testing and is allowed to cool down for a set time interval. This helps to arrive at a cooling rate of the first core from the heating of the second core.

1. *Creating Thermal Database 1 (TD1): TD1* deals with the temperature profiles of cores due to their own leakage power consumption alone. The procedure to calculate the corresponding peak temperature values is noted in Procedure *Compute_TD1*. It creates a power profile for a core assuming that the core operates for 2000 clock cycles and consumes leakage power only. For this, the inputs to the core can be held at some constant values, so that the core circuit does not undergo any transition. Power values for all other cores are taken to be zero. This power profile (leakage power for core i and zero for others) is fed to a thermal simulator, along with the SoC floorplan. The temperature rise for core i is noted. This quantity $T_{leakage\_i}$ is stored in *TD1* for core i.

Procedure Compute_TD1

Input: SoC floorplan
Output: $T_{leakage\_i}$ for core i
Begin

    For each core i do

        Compute leakage power $P_{leakage\_i}$ for core i for 2000 clock cycles;
        Set power of core i as $P_{leakage\_i}$ and zero for other cores;
        Feed power profile and SoC floorplan to a thermal simulator;

Calculate maximum temperature $T_{leakage\_i}$
from transient and steady state responses
for core i;

Enter $T_{leakage\_i}$ into *TD1*;

End.

2. *Creating Thermal Database 2 (TD2)*: This procedure calculates
the rise in peak temperature of a core under two conditions.
The first situation assumes that only core i is being tested.
Thus, core i consumes dynamic power while all other cores
are in idle mode, consuming only leakage power. A power
profile is created by considering core i consuming dynamic
power for a number of clock cycles equal to its test time; for
other cores, leakage power is taken for the same number of
cycles. This power profile, along with the SoC floorplan is fed
to a thermal simulator to see the peak rise in the temperature
of core i. This quantity, marked as T_{self_i} is entered in *TD2*
at (i,i). The next situation takes into consideration the case
when another core (core j) is being tested in parallel with core
i. Power profiles for the cores are created by assuming that
cores i and j consume their corresponding dynamic power for
the duration equal to the test time of core i. For other cores,
power values are taken to be equal to their leakage power
consumptions with a duration equal to the test time of core
i. This power profile and SoC floorplan are fed to a thermal
simulator to note the maximum temperature reached for
core i. Let this quantity be $T_{parallel\_ij}$. The additional rise in
the peak temperature of core i due to its parallel testing with
core j is computed in T_{extra_ij} as

$$T_{extra\_ij} = T_{parallel\_ij} - T_{self\_i}$$

This T_{extra_ij} is put into TD2 at location (i,j). Procedure
Compute_TD2 details the steps involved in the computation.

Procedure Compute_TD2

Input: SoC floorplan

Output: T_{self_i} for core i and T_{extra_ij} for each pair of cores i and j

Begin

For each core i do

Set power of core i as $P_{dynamic\_i}$ for test time of the core;

Set power of other cores as their leakage power values for the same duration;

Feed power profile and SoC floorplan to a thermal simulator;

Calculate maximum temperature T_{self_i} from transient and steady state responses for core i;

Enter T_{self_i} into TD2 at location (i,i);

For each core j ($j \neq i$) do

Set power of core i as $P_{dynamic\_i}$ for test time of core i;

Set power of core j as $P_{dynamic\_j}$ for test time of core i;

Set power of other cores as their leakage power values for the same duration;

Feed power profile and SoC floorplan to a thermal simulator;

Calculate maximum temperature $T_{parallel\_ij}$ from transient and steady state responses for core i;

$$T_{extra\_ij} = T_{parallel\_ij} - T_{self\_i}$$

Enter T_{extra_ij} into TD2 at location (i,j);

End.

3. *Creating Thermal Database 3 (TD3)*: TD3 notes the effect on temperature of core i due to a prescheduled core j finishing

testing just before the beginning of testing of core i. A power profile is created in which core j consumes its dynamic power for a duration equal to its test time, while all other cores (including i) consume leakage power for the same duration. This power profile is fed to a thermal simulator and the temperature of core i at the end of test time of core j is noted. This temperature $T_{preschedule1\_ij}$ is put into the position (i,j) of table *TD3*. Procedure *Compute_TD3* details the process.

Procedure Compute_TD3

Input: SoC floorplan
Output: $T_{preschedule1\_ij}$ for each pair of cores i and j
Begin

    For each pair i, j of cores do

        Set power of core j as $P_{dynamic\_j}$ for test time of core j;

        Set power of other cores as their leakage power values for the total duration;

        Feed power profile and SoC floorplan to a thermal simulator;

        Find temperature $T_{preschedule1\_ij}$ for core i;

        Enter $T_{preschedule1\_ij}$ into *TD3* at location (i,j);

End.

4. *Creating Thermal Database 4 (TD4)*: TD4 allows a cooling period between the scheduling of two cores. The corresponding cooling rate can be calculated from observation. For each pair of cores i and j, such that core i is scheduled for testing after core j has finished, the cooling rate is computed as follows. A power profile is created in which core j consumes its dynamic power for a duration equal to its test time, followed by a cooling period of 2000 clock cycles. During this time, core j consumes only leakage power. For all other cores (including i), power values are taken to be equal to their leakage power for the total duration (test time of core j + 2000). The power

profile is fed to a thermal simulator to get the temperature of core i at the end of this time duration. Let the corresponding temperature be $T_{preschedule2\_ij}$. This is stored in $TD4$ at location (i,j). The cooling rate of core i with respect to core j, CR_{ij}, can be calculated as follows:

$$CR_{ij} = \frac{T_{preschedule1\_ij} - T_{preschedule2\_ij}}{2000}$$

This cooling rate, multiplied by any time gap TG between the end time of core j and start time of core i can give the net increase in the initial temperature of core i due to the scheduling of core j before i (NAI_{ij}).

$$NAI_{ij} = T_{preschedule1\_ij} - (CR_{ij} \times TG)$$

Procedure *Compute_TD4* shows the detailed steps to compute $T_{preschedule2\_ij}$.

Procedure Compute_TD4

Input: SoC floorplan
Output: $T_{preschedule2\_ij}$ for each pair of cores i and j
Begin

    For each pair i, j of cores do

        Set power of core j as $P_{dynamic\_j}$ for test time of core j;

        Set power of core j as $P_{leakage\_j}$ for 2000 clock cycles;

        Set power of other cores as their leakage power values for the total duration;

        Feed power profile and SoC floorplan to a thermal simulator;

        Find temperature $T_{preschedule2\_ij}$ for core i;

        Enter $T_{preschedule2\_ij}$ into $TD4$ at location (i,j);

End.

4.4.2 Phase II

At the end of phase I, we have a set of test rectangles corresponding to each core. For each such rectangle, the corresponding test time, power, and temperature values have been computed. The thermal parameters have been put into thermal databases *TD1* through *TD4*. The next step is to select one test rectangle for each core and schedule the rectangles satisfying the power and temperature limits of the SoC. Overall test time of the SoC is determined in the process. In the following, test rectangles are selected using a PSO technique in which each particle gives a selection of test rectangles for the cores. Scheduling has been carried out using a heuristic technique that takes into consideration power and thermal issues.

4.4.2.1 PSO Formulation

Let the number of cores in the SoC be N and the maximum number of rectangles for any core in the SoC be M. Let, $B = \lceil \log_2 M \rceil$. For the PSO formulation, a particle consists of $N \times B$ number of bits. Each group of B bits corresponds to the rectangle selection for a core. The first B bits identify the test rectangle selected for the first core, second B bits for the second core, and so on. For example, for $N = 4$ and $B = 4$, a particle "1001001010001101" corresponds to the selection of rectangles 9, 2, 8, and 13 for the cores 1, 2, 3, and 4, respectively. For the initial generation, particles are created randomly to constitute the population. However, care has to be taken to ensure that the indices generated for a core are always within the total number of test rectangles of it.

As noted in Chapter 2, in a PSO formulation, evaluation of a particle is guided by three parameters—its own (local) intelligence, the swarm (global) intelligence, and the inertia factor. A particle remembers its history about its best structure over the generations. This is called the local best (*pbest*) of the particle. In a particular generation, the particle with the best fitness value is called the global best (*gbest*) of the generation. For the initial generation, *pbest* of each particle is initialized to itself, while the *gbest* for the population is set to the best particle. The fitness calculation procedure has been detailed later. For the creation of successive generations, new particles are evolved using the *replace* operator, noted next.

The *replace* operator attempts to align a particle with its own local best, *pbest* and the global best of the generation, *gbest*; both with some probabilities. For this, the *replace* operator is applied at each bit position of the particle. For bit position i of a particle, the bit is replaced by the corresponding bit of the *pbest* particle with probability α. Next, on the modified particle, the *replace* operator is applied to align it with *gbest*, following a similar procedure with probability β. The values of α and β are user tunable. The experimental results reported in this chapter use a value of 0.1 for both α and β. After the replace operator has been utilized to modify a particle, guided by *pbest* and *gbest* particles, a consistency check is performed to see that the resulting rectangle indices are valid for all cores. If the new rectangle number for a core in the resulting particle becomes larger than the total number of rectangles available for the core, the rectangle number is reverted back to its value in the original particle.

4.4.2.2 Particle Fitness Calculation

A particle in PSO suggests the test rectangles selected for each core in the SoC. Fitness of the particle is evaluated by performing a scheduling of these rectangles. A schedule gives the total time needed for testing all of the cores in the SoC. This TAT is taken as the fitness of the particle. The scheduling heuristic needs to take care of the following: availability of enough TAM wires needed by a test rectangle and the adherence to the maximum power and temperature budgets for the SoC.

The heuristic scheduling algorithm takes as input the rectangle set corresponding to the particle, the maximum TAM width W_{max}, the maximum power limit P_{max}, and the maximum allowable temperature T_{max}. It performs a scheduling of the rectangles ensuring the constraints that at no instant of time, the total TAM width requirement exceeds W_{max}, the instantaneous power value exceeds P_{max}, or the maximum temperature of any core exceeds T_{max}. The algorithm maintains the following data structures to arrive at scheduling decisions of cores.

1. *Break_Point_List (BP)*: A set of time instants at which the power requirement of the schedule has changed from its value in the previous instant. The next core can be scheduled at any of the breakpoints, $bp_k \in BP$.

2. *Available_TAM_Width_Info (ATW)*: A set with cardinality the same as *BP*. The value atw_k is equal to the total free TAM width available at breakpoint instant bp_k.

3. *Power_Tracker (PT)*: This is also a set with cardinality the same as that of *BP*. It holds the total power consumed by the already scheduled cores at the corresponding breakpoint instant.

4. *Temperature_Tracker (TT)*: It contains the temperature values of each core during test scheduling.

A *Power_Violation_Checker* (PVC) checks for power violations in the schedule. At any point in the schedule, if an attempt is made by the algorithm to schedule a core but it does not respect *PVC* (that is, the total power budget of SoC gets violated by scheduling the core at that time), it does not get scheduled at that point. Similarly, a *Thermal_Violation_Checker* (TVC) checks whether any core is violating the thermal budget or not. *TVC* checks the scheduling position of a core and uses the superposition principle to add different thermal values from *TD1, TD2, TD3,* and *TD4* corresponding to that core, to calculate the exact temperature of the core. Suppose that a core starts its testing at time $t = 0$. Thus, there will be no leakage power consumption of the core before it starts testing. Naturally, the increase in the initial temperature should not be considered for this core. However, the effect of increase in initial temperature has to be considered for the cores that remain idle for some time and thus consume leakage power. Similarly, when a new core gets scheduled, the temperature profiles of the already scheduled cores also get modified because of the newly scheduled core. *TVC* checks all these possible conditions and decides whether or not the new core can be scheduled at a breakpoint. The temperature calculation procedure has been noted in the procedure *Compute_TVC* [3].

Procedure Compute_TVC

Input: Thermal databases *TD1, TD2, TD3, TD4*;
      Core C_i to be scheduled;
      bp_k, the breakpoint at which scheduling decision has to be taken;
Output: "valid" if C_i can be scheduled at bp_k, "invalid" otherwise;
Begin
      // Calculate temperature T_i of C_i
      If (idle time of C_i is more than 1000 clock cycles) then
            Add $T_{leakage\_i}$ from *TD1* to T_i;
            Check thermal violation for C_i;
            If (thermal violation detected)
                  Return "invalid";
      Add T_{self_i} from *TD2* to T_i;
      Check thermal violation for C_i;
            If (thermal violation detected)
                  Return "invalid";
      For each core C_j ($j \neq i$) already scheduled do
            Check parallelism between C_i and C_j
            If (parallel)
                  Add T_{extra_ij} to T_i;
                  Add T_{extra_ji} to T_j;
                  Check thermal violation for C_i and C_j;
                  If (thermal violation detected)
                          Return "invalid";

Else   // No parallelism, cooling to be considered
Calculate CR_{ij};
Find scheduling time gap (TG) between C_j and C_i;
$T_{net\_increase} = T_{preschedule1\_ij} - (CR_{ij} \times TG)$;
Add $T_{net\_increase}$ to T_i;
Check thermal violation if C_i;
If (thermal violation detected)
    Return "invalid";
Return "valid";

End.

To make the schedule compact, it is necessary to utilize any TAM resource remaining idle at any point in time. Also, an attempt has to be made to utilize the power and temperature budgets completely. To accomplish this, for the breakpoint bp_k, the algorithm scans the unscheduled rectangle list to look for the largest rectangle that can be scheduled at bp_k. If none of the rectangles meet the requirements, the algorithm advances to the next breakpoint. Power and temperature validations are carried out to ensure the satisfaction of power and temperature limits at every breakpoint until the end of schedule for the current core. When rectangles corresponding to all of the cores have been scheduled, the maximum of the test-completion times of individual cores gives the total TAT for the SoC. The rectangle scheduling process has been noted in Algorithm *Schedule_Rectangles*.

Algorithm Schedule_Rectangles
Input: List of rectangles for cores; TAM width W_{max}; Power limit P_{max}; Temperature limit T_{max};
Output: Power and temperature valid schedule for all cores;
Begin
    While (all cores are not scheduled) do
        $bp_k \leftarrow$ Breakpoint with the minimum time value;

$atw_k \leftarrow$ Available TAM resource at bp_k;

While ($atw_k > 0$) do

$Area_{max} \leftarrow 0$;

For all unscheduled cores C_i do

$w_{ij} \leftarrow$ width of the test rectangle of C_i;

$P_{ij} \leftarrow$ corresponding power consumption

If ($w_{ij} \leq atw_k$) then

If P_{ij} respects PVC then

If C_i and all other previously scheduled cores respect *TVC* then

Calculate $Area_{ij} = w_{ij} \times testtime_{ij}$;

If $Area_{ij} > Area_{max}$ then

$Area_{max} \leftarrow Area_{ij}$;

$C_{i\_max} \leftarrow C_i$;

Select C_{i_max} for scheduling at bp_k;

Update *BP*;

Update *ATW*;

Update *PT* with P_{ij};

Update *TT* with current temperature values of all cores;

$Start_{C\_i} = bp_k$;

$End_{C\_i} = bp_k + testtime_{ij}$;

Mark C_i as scheduled;

Remove bp_k and atw_k from *BP* and *ATW* respectively;

End.

4.5 EXPERIMENTAL RESULTS

For the sake of experimentation, two SoCs named *k10* and *k25* have been created with 10 and 25 cores respectively, taken from the ITC99, ISCAS89, and ISCAS85 circuits. Individual circuits from the benchmark suite have been converted into cores as follows:

1. Each circuit description in Verilog format has been taken as input and mapped to a Faraday 90 nm standard cell library.

2. All flip-flops are replaced by scan flip-flops.

3. Multiple scan chains are inserted to reduce the test application time.

4. Dynamic and leakage power values are extracted using power estimation tools.

5. Area of each core is calculated.

6. Floorplan for an SoC is obtained using floorplanning tools. As cores have different sizes, empty spaces get created into the SoC floorplan. These empty spaces are packed with zero power boxes to help in the thermal simulation process.

7. Test vectors are generated for the individual cores using ATPG tools.

The composition of the two SoCs has been shown in Table 4.1.

The results of the PSO-based thermal-aware test scheduling policy are shown in Tables 4.2 and 4.3 for the SoC benchmarks *k10* and *k25*, respectively. The population size for PSO has been taken to be 2000 and 3000 particles for *k10* and *k25*, respectively. PSO evolution terminates if there is no improvement in solution over 1000 and 2000 generations for *k10* and *k25*, respectively. Results are noted for $W_{max} = 64$ and 56, for both the SoCs along with variations in P_{max} and T_{max} values. It may be observed that with increase in P_{max} and T_{max} values, the TAT for the SoCs gets reduced. This happens as with the relaxation in power and thermal constraints, TAT can be minimized further. One can choose P_{max} and T_{max} values, according to the requirement, without violating the tolerance limits for the SoC, to obtain the best solution. It may be noted that, in some cases, increase in the T_{max} value does not reduce the TAT value further. This is due to the fact that the SoC

TABLE 4.1 SoCs *k10* and *k25*

Circuit Name	Dynamic Power (mW)	Leakage Power (μw)	Area (μm²)	No. of Test Vectors	No. of Scan Chains	Max. Length of Scan Chain	No. of Copies in SoC	
							k10	*k25*
s38584	9.1529	315.5739	0.129	146	32	45	2	2
s38417	4.6092	311.4597	0.134	100	32	55	2	0
s35932	10.2743	282.4460	0.118	64	32	54	0	2
s15850	3.0535	128.6255	0.054	131	16	34	1	0
s13207	1.9165	116.4896	0.049	273	16	41	0	1
s9234	1.2388	71.0824	0.029	147	4	54	1	2
s5378	0.7293	35.0989	0.016	124	4	46	0	2
b22	4.3550	218.5332	0.093	1501	46	15	0	1
b21	2.5584	143.3848	0.061	711	16	31	1	2
b17	3.6796	346.3996	0.161	1477	46	31	0	3
b15	1.1875	120.6018	0.058	457	16	29	2	2
b14	1.3280	68.1279	0.029	737	4	62	1	1
c7552	3.5920	38.4599	0.015	219	0	0	0	3
c5315	2.2925	27.8943	0.011	123	0	0	0	2
c1908	0.9554	7.5269	0.003	119	0	0	0	1
c499	0.0040	1.1222	0.0006	52	0	0	0	1

has already reached its saturation TAT value in that temperature range. Hence, variation in temperature does not have any further effect on the TAT in that range.

4.6 SUMMARY

In this chapter, a superposition principle-based fast thermal model has been presented. The model is capable of estimating temperature with good accuracy. A PSO-based formulation has been developed to select test rectangles for the cores in SoC. A heuristic scheduling strategy has been reported for these selected test rectangles which maintains the power and thermal limits specified for the SoC. Experimental results have been reported on SoCs constructed using benchmark combinational and sequential circuits. A good trade-off can be achieved between test application time, permissible power, and temperature limits.

TABLE 4.2   Variation in TAT with Power and Temperature Limits for SoC *k10*

	$T_{max}(^\circ C)$	TAT for P_{max} (in Watts) Equal to				
		1.3	1.5	2.0	5.0	∞, $T_{max} = \infty$
$W_{max} = 64$	72	43154	42494	42365	42365	
	80	41424	41388	41347	40924	
	85	39349	37875	37875	37875	
	90	39349	37875	37875	37875	34021
	95	39349	36984	36110	35356	
	100	39349	35880	34810	34459	
	105	39349	35865	34810	34072	
	110	39349	35732	34268	34021	
$W_{max} = 56$	72	50285	50285	50285	49010	
	80	43530	42424	42424	42424	
	85	43530	39672	39672	39672	
	90	43530	39672	39672	39672	38933
	95	42839	39672	39672	39672	
	100	42300	39672	39672	39672	
	105	42300	39672	39672	38936	
	110	42300	39672	38936	38936	

TABLE 4.3   Variation in TAT with Power and Temperature Limits for SoC *k25*

	$T_{max}(^\circ C)$	TAT for P_{max} (in Watts) Equal to				
		2.7	3.0	3.5	4.5	∞, $T_{max} = \infty$
$W_{max} = 64$	82	171142	169876	169869	169684	
	90	170578	169562	169146	169101	
	95	169740	169101	169101	168908	167971
	100	169513	169101	168982	168778	
	105	169090	168982	168904	168698	
	110	168698	168602	168602	168602	
$W_{max} = 56$	82	196567	196567	193613	192163	
	90	195710	193202	192494	192050	
	95	193074	192400	192163	191276	190231
	100	192494	191894	191276	191276	
	105	191814	191894	191276	191276	
	110	191814	191894	191276	191276	

REFERENCES

1. V. Iyengar, K. Chakrabarty, and E.J. Marinissen, "Test Wrapper and Test Access Mechanism Co-optimization for System-on-Chip", *Journal of Electronic Testing*, vol. 18, No. 2, pp. 213–230, 2002.
2. R. Karmakar, S. Chattopadhyay, "Window Based Peak Power Model and Particle Swarm Optimization Guided 3-Dimensional Bin Packing for SoC Test Scheduling", *Integration, the VLSI Journal*, vol. 50, pp. 61–73, 2015.
3. R. Karmakar, S. Chattopadhyay, "Thermal-Safe Schedule Generation for System-on-Chip Testing", *International Conference on VLSI Design*, 2016.

CHAPTER 5

Network-on-Chip Testing

5.1 INTRODUCTION

In SoC-based systems, designers integrate various cores taken from different vendors into a single chip. Vendors provide not only the cores, but also the set of test patterns to test the cores. The test set provided by the vendors guarantees proper testing of the core alone, not the entire system chip. The test sets provided by the vendors of individual cores need to be applied to the input lines of the cores and the output responses from the cores are to be collected at the ATE outside the chip. This poses a challenge as the input-output lines of the integrated cores are not directly accessible from the system input-output pins. In an SoC environment (noted in Chapter 4), such problems are solved by providing an extra dedicated *test-access mechanism* (TAM) for the chip. TAM is accessible from the input-output pins of the system. The core pins are accessible from the TAM lines. In test mode, test patterns are carried by the TAM and applied to the core. The generated test responses from the core output lines are carried to the output port of the system through these TAM lines. The overall TAM architecture is often divided into a number of busses to carry out the testing of cores in parallel. Moreover, some of the cores may support preemption during the test process, allowing the test session for the core to be interleaved with the testing of others. Many other cores may not support the preemption mechanism, implying that all the test patterns for the core are to be applied in a single test session for the core.

In a *network-on-chip* (NoC) environment, usage of extra TAM is not advisable. An NoC-based system chip contains within it a network for communication between the cores. The network consists of a set of simple routers interconnected in some topology (such as, mesh, tree etc.) via dedicated links. Individual cores are connected to the local ports of routers. All electrical signal exchanges between cores are replaced by messages passing through the underlying network. This solves the communication bottleneck problem of a bus-based SoC design paradigm, as the bus architecture limits the degree of parallel communication that can be achieved in the system, restricting the system throughput. For testing an NoC-based system, such a network can be reused to carry the test information for the cores. In an NoC-based system, some dedicated cores are used as input/output (I/O) ports/pairs for testing the entire system. The *automatic test equipment* (ATE) connects to the system through these ports to supply test patterns and collect responses. Figure 5.1 shows such a configuration for the benchmark SoC *d695*. It has two I/O pairs indicated by the incoming and outgoing arrows in the figure. A channel of ATE is attached to each I/O pair. To test a core, it has to be attached to an I/O pair. For example, in Figure 5.1, cores 1 and 9 form one I/O pair. The other I/O pair consists of cores 2 and 6. If core 7 has to be tested by the I/O pair (1,9), core 1 (source) will feed the test patterns to core 7. The test and response packets will move through the NoC following a deadlock-free deterministic routing (such as *xy*-routing for mesh). Test responses are collected in core 9 (sink). Test packets and responses move in a pipelined fashion through this path.

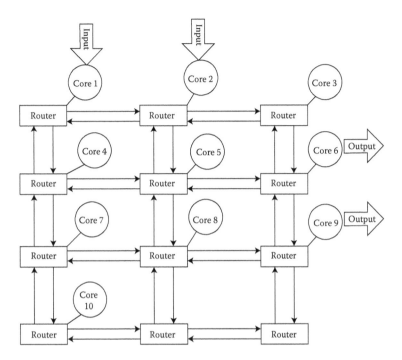

FIGURE 5.1   NoC for SoC *d695* with input-output pairs.

Similarly, core 5 is assigned to the I/O pair (2,6). A packet consists of *flits*. A flit consists of bits delivered to a router from its predecessor per unit time. If the link width is w bits, the flit size is also w bits. The flits are unpacked in one cycle when they arrive at a particular core.

The rest of the chapter is organized as follows. Section 5.2 presents the problem statement. Section 5.3 enumerates the test time calculation procedure for NoC. Section 5.4 presents a model for temperature estimation of the NoC. Section 5.5 elaborates a PSO formulation for the preemptive test-scheduling problem. Section 5.6 shows some augmentations to the basic PSO that can be utilized for achieving better results. Section 5.7 presents the overall augmented PSO algorithm. Section 5.8 details some experimental results. Section 5.9 summarizes the chapter.

5.2 PROBLEM STATEMENT

A core supporting preemption can be scheduled in either a preemptive or a non-preemptive fashion. Preemptive test scheduling approaches [1] often result in lower test time than the non-preemptive ones, as in the former case, tests can be preempted to avoid resource conflicts. Resource conflicts occur when two or more core tests require one or more common links to transport-test patterns or responses. As a result, idle times need to be inserted in the overall test schedule to avoid such conflicts. In preemptive test scheduling, these idle times can be utilized to test other cores. The preemptive strategy also does not significantly increase the ATE's control complexity. It stores the test patterns in the same way as in a non-preemptive strategy, except that the sequences of test patterns are recorded in the ATE memory. Therefore, a judicious mix of sequences, testing preemptive, and non-preemptive cores can have better flexibility in utilizing test times, which would otherwise remain idle because of resource conflicts. Furthermore, to reduce test time, test engineers often perform parallel testing of cores. This may result in power and temperature violations for the chip. The cores may also be tested at different frequencies. Thus, the overall test-scheduling problem can be stated as follows:

> Identify distribution of clock frequencies, I/O pairs, and test start times for each core such that the overall NoC test time is minimized, satisfying chip power constraint and ensuring thermal safety of the system.

The input to the test-scheduling problem consists of the following.

- A set $C = \{C_i, 1 \leq i \leq N_c\}$ of N_c cores in the NoC with test time T_i for core C_i.

- The test property of each core—preemptive or non-preemptive.

- A set $F = \{F_i: f/m \leq i \leq m \times f\}$ of $(2m - 1)$ frequencies that can be generated from the tester clock frequency f, where m

is the rate for faster and slower clocks. In this chapter, the maximum and the minimum clock rates are $2f$ and $f/2$, respectively, that is, $m = 2$. Therefore, the available frequency list is $F = \{f/2, f, 2f\}$.

- A set $IO = \{IO_i : 1 \leq i \leq k\}$ of k cores marked as input-output pairs.

5.3 TEST TIME OF NOC

In preemptive test scheduling, test packets are distributed in several chunks. Test time for a chunk can be calculated as follows. Let p denote the number of test patterns in the chunk. Let, the maximum scan-chain length be l. Because only 1 bit can be shifted into the scan chains at a time, from each flit, each scan chain can get only a single bit. Thus, the test packet consists of l such flits. Let the distance (in number of hops) from test source to the core under test and from there to the sink core be denoted as $h_{s \to c}$ and $h_{c \to k}$, respectively. As each flit is unpacked in one clock cycle and flits move in a pipelined fashion, the time to send a test packet from source to the core plus the time to collect response in the sink is given by

$$h_{s \to c} + (l-1) + 1 + h_{c \to k} + (l-1).$$

During testing, consecutive flits are applied to the core. So the minimum inter-arrival time between consecutive flits is given by *Max{time required to shift-in a flit into the core, time required to shift-out a response from core} + unit cycle to unpack the flit + time taken to generate the response.* Therefore, the entire test time T for a particular chunk is given by

$$T = h_{s \to c} + (l-1) + 1 + \left[1 + Max\{h_{s \to c} + (l-1), h_{c \to k} + (l-1)\}\right]$$
$$\times (p-1) + h_{c \to k} + (l-1)$$
$$= \left[1 + Max\{h_{s \to c}, h_{c \to k}\} + (l-1)\right] \times p + \left[Min\{h_{s \to c}, h_{c \to k}\} + (l-1)\right]$$

A core can be operated at different clock rates and these clock rates are some multiples of the basic tester clock. Faster ($2f$) and slower ($f/2$) clocks are generated from the tester clock (f), which are double ($m = 2$) and half ($m = 1/2$) of the tester clock, respectively. Thus, the overall test time for a chunk is given by $T' = T/m$.

5.4 PEAK TEMPERATURE OF NOC

Temperature of a tile/IP-block depends on its power consumption and its position in the NoC floorplan. As noted in Chapter 1, typically, circuits are packaged with a configuration such that the die is put against the heat spreader. The spreader is placed against the heat sink, which can be cooled by a fan. Simple and compact thermal models calculate the temperature of each IP block by considering the heat dissipation within the block and also the effect of heat transfer between the IP blocks, based on an RC model. The RC model considers three verticals: conductive layers for the die, heat spreader, and heat sink. It also takes care of the fourth vertical—convective layer for the sink to air interface. Heat flow paths from four sides of the spreader and inner and outer portions of the heat sink are also taken into consideration. The thermal resistance $R_{i,j}^{th}$ of IP block IP_i with respect to IP_j is defined as

$$R_{i,j}^{th} = \Delta T_{i,j} / \Delta P_j$$

where, $\Delta T_{i,j}$ is the increment in the temperature of IP_i due to ΔP_j power dissipation at IP_j. The corresponding thermal resistance matrix is given by R^{th}. If P and T denote the power and temperature vectors, then $T = R^{th} \times P$. If the total number of tiles in the NoC is q, the maximum temperature T_{peak} is equal to the maximum of the values $T_1, T_2, \dots T_q$. Next, A PSO formulation for the preemptive test-scheduling problem is presented.

5.5 PSO FORMULATION FOR PREEMPTIVE TEST SCHEDULING

As noted in the previous chapters, in any PSO formulation, solutions are represented as particles. In this problem, a particle represents a candidate test schedule of cores [2]. It has four components—core part, I/O part, frequency part, and preemption part. The core part corresponds to a permutation of cores C_1 to C_n. It represents an order in which the test scheduler will pick up the cores for probable scheduling and assign it to the corresponding time slot. The next part is an array of I/O pairs. It is of same size as the core part. If k I/O pairs are available, each entry contains an integer between 1 and k. The third part also contains an array of size equal to the number of cores. If m frequencies are available, each entry in this array contains one of the m possible frequencies. Each preemption part entry is a real number between 0 and 1. For non-preemptive cores, the entry is always 1. For a preemptive core, if the corresponding entry is x and the core has a total of p test patterns, $x \times p$ number of patterns will be tested as a single block. Thus, testing of the entire core gets distributed over a number of blocks, each block being scheduled independent of others. However, because the same I/O pair is attached to the core in each test session, the same set of routing resources will be utilized for them.

To compute the fitness of a particle, the scheduler first determines the overall test time for the cores. For this, the scheduler starts with the first core from the core part. Throughout the entire test of a core, proper NoC resources, such as links and routers, remain reserved. During preemption, reserved resources can be released for testing of other cores. The resources needed for testing a core are decided by the I/O pair, core-under-test, and the routing algorithm. At a certain time instant, suppose that the scheduler has scheduled up to the ith core in the core part of a particle. Now, scheduling time for $(i + 1)$th core will be decided by consulting the corresponding I/O pair and frequency parts.

If the corresponding I/O pair is j, the next available slot for j is determined by the block $x \times p$, where x represents the preemption coefficient of the $(i + 1)$th core and p denotes the total test packets for the core, so that proper resources are available for the testing of the $(i + 1)$th core. The scheduler schedules the cores until there are no test packets left for any core. The overall test time of the entire schedule is computed from the highest time of any of the I/O pairs. The test time for each chunk is computed by the formulation shown in Section 5.3.

Once the test schedule has been determined, the next task is to identify the peak temperature of the NoC resulting from the testing of the cores [3]. The schedule gives the power consumed by different NoC tiles at various time instants. This gives the power profile of the NoC during the test session. The methodology enumerated in Section 5.4 can be utilized to obtain the peak temperature of the NoC tiles. The overall fitness of a particle is computed as follows:

$$Fitness = W \times \frac{T_{peak}}{\alpha} + (1 - W) \times \frac{Test\ time}{\beta} \tag{5.1}$$

where T_{peak} is the peak temperature and *Test time* is the overall test time for the schedule corresponding to the particle. W is a weighing factor to balance between the optimization of peak temperature and test time. W is in the range 0 to 1. $W = 1$ optimizes the peak temperature alone, while $W = 0$ optimizes only the test time. As test time and temperature values belong to different ranges, normalization of these two metrics has been done by assuming their worst case scenarios. To set the value of α, test scheduling is performed for all cores, assuming the availability of only one I/O pair. For each candidate I/O pair available in the NoC, this evaluation is carried out. The maximum among these gives the normalization factor α. The factor β is set to the peak temperature reached, assuming that all cores are being tested parallel.

As noted in Chapter 2, PSO evolution can be continuous or discrete. A continuous evolution process computes the velocity of the particle in each direction and changes the position of the particle determined by these velocity components. On the other hand, discrete PSO formulation changes the particle position by operators like *swap* or *replace*. In the preemptive NoC test-scheduling policy, both the continuous and discrete PSO versions have been utilized. Among the four parts of a particle, the core, I/O pair, and frequency parts evolve using the principle of discrete PSO, while the preemption part evolves via continuous PSO policy.

5.6 AUGMENTATION TO THE BASIC PSO

To improve the quality of the solution produced by the basic PSO formulation, enumerated in Section 5.5, an augmentation can be made to run the PSO multiple times, called *Multi-PSO*. In any population-based search technique, exploration and exploitation are the two properties that can be used to control the quality of the solution. Similar techniques can also be incorporated in PSO. In the exploration phase, different regions of the search space are explored, whereas the exploitation process checks for local optima around the globally explored points. In the initial portion of a PSO run, it performs more of an exploration. However, through the evolution process, the particles start converging, making more of an exploitation. To balance between the exploration and exploitation in a multiple swarm-based optimization, the *locust swarm* policy uses a *devor and move on* strategy. In this strategy, if a sub-swarm has found local optima, a set of scouts are deployed to explore the new potential regions. Furthermore, the scouts are guided by intelligence accumulated by their earlier sub-swarm. Similar strategy can be utilized to guide the NoC test scheduling PSO as well.

In the suggested augmentation, PSO can be run several times to improve upon the global best solution. Suppose that at the end of nth run of PSO, the local best for the kth particle is $pbest_n^k$ and the global best is $gbest_n$. In the $(n + 1)$th pass of the PSO, it starts with a new set of particles. However, the local and the global best information are transferred from the nth to the $(n + 1)$th PSO. The maximum number of PSO runs can be controlled as follows:

- A user-defined value for the maximum number of PSO runs. The results reported in this chapter works with a value of 2000 PSO runs.

- The global best fitness does not change in the last 100 PSO runs.

5.7 OVERALL ALGORITHM

The complete PSO engine has been presented in Algorithm *NoC_ Schedule* [4]. The algorithm generates *NPart* random particles to constitute the initial generation of population. The local best of individual particles is set to the particle itself. The global best is set to the particle with the fittest configuration. In the algorithm, *MGEN* is the maximum number of generations for which individual PSO may run. The total number of PSO runs has been represented by *MPSO*.

After generating the initial configuration and finding the global best configuration of the first generation, the particles are evolved using *UpdatePart()*. Each particle evolves by sharing the experience of its local as well as the global best of the generation with some probabilities. *Swap* sequences are generated for core, I/O pair and frequency parts of the particle. The preemption part evolves via continuous PSO evolution policy. After each generation, the PSO engine checks the *BestFitness* value. The current generation number is reset to zero if the PSO gets better solution than the earlier one. Otherwise, it keeps a count on the generations. Completion of a single PSO run is determined by either this count reaching a user-defined value or the PSO has been run for a predetermined number of generations. For multiple PSO runs, the PSO engine creates new particles by assigning random configurations, whereas, the local best configurations for the particles are passed from the previous PSO run. The PSO engine stops when multiple PSO run counts reach the value *MPSO*.

Algorithm NoC_Schedule

Input: NoC topology, core mapping, test patterns
Output: Test schedule for cores
Begin
Step 1: Set values for *MGEN, MPSO, NPart, W*.
Step 2: *BestFitness* ← ∞.
Step 3: For *m* = 0 to *MPSO* do
    Step 3.1: *BeforeBestFitness* ← ∞; *gen* ← 0;
    Step 3.2: While *gen* < *MGEN* do
        For *p* = 0 to *NPart* do
            If *gen* = 0 and *m* = 0 then
                // *Initialize population*
                $Particle_p$ ← *Random()*;
                $Particle_p^{pbest}$ ← $Particle_p$;
            Else
                // *Update particle*
                UpdatePart(p);
                Compute fitness of $Particle_p$;
                UpdateBest($Particle_p^{pbest}$, $Particle_p$);
            End;
            Compute global best particle.
            Update the *BestFitness* as global best and reset *gen* to zero.
            gen ← *gen* + 1.
        End.
    Step 3.3: for *p* = 0 to *NPart* do
        $Particle_p$ ← *Random()*;
        CopyLocalBest ($Particle_p^{pbest}$);
        End;
    End;
Step 4: Output the best particle.
End.

5.8 EXPERIMENTAL RESULTS

In this section experimental results have been presented on *ITC'02* SoC benchmarks—*d695, p22810*, and *p93791*. For each, the floorplan has been generated manually. The core power values have been generated randomly, maintaining the configuration noted in Table 5.1. The router and link power values have been generated using the tool *Orion*. Individual tiles in the NoC are of dimension 2.5 mm × 2.5 mm. The tool HotSpot has been used with its default parameter setting and the ambient temperature of 318.15°K for thermal simulation.

5.8.1 Effect of Augmentation to the Basic PSO

Table 5.2 shows the effect of augmentation, suggested in Section 5.6 to the basic PSO formulation. PSO formulation with and without the proposed augmentation have been run on SoC benchmarks *p22810* and *p93791*. The column marked "Basic PSO" notes the results of the original PSO formulation, while the column marked "Multiple PSO" contains the results of the augmented PSO. It can be observed that the augmented PSO reduces test time significantly for the two SoCs. In both the cases, four I/O pairs have been assumed for supplying test patterns to the NoC.

TABLE 5.1   Power Ranges for Cores of SoCs

SoC	No. of Cores	Power Values for Cores in mW		
		Maximum	Minimum	Average
d695	10	4003	255	2014.10
p22810	28	9128	176	4302.29
p93791	32	9128	554	4674.38

TABLE 5.2   Test-time Comparison Between Basic and Augmented PSO

SoC	I/O Pairs	Test time	
		Basic PSO	Multiple PSO
p22810	4	88167	85453
p93791	4	100295	85967

TABLE 5.3   Test-time Comparison between Preemptive
and Non-preemptive Approaches

		Test time values	
SoC	I/O Pairs	Preemptive PSO	Non-preemptive PSO
d695	2	9807	9807
	3	6933	6933
	4	5272	5272
p22810	2	145098	145563
	3	98167	104462
	4	85453	97644
p93791	2	159188	160773
	3	110338	121089
	4	85967	114317

5.8.2 Preemptive vs. Non-preemptive Scheduling

To compare the test time between the preemptive and non-preemptive versions of the NoC test-scheduling problem, the PSO formulation noted in Section 5.7 has been modified to ignore the preemption part of the particles. Table 5.3 notes the comparative results for ITC benchmarks with different number of I/O pairs. Test time decreases with the increase in the number of I/O pairs. The preemptive test scheduling performs better than the non-preemptive version, particularly for larger benchmarks.

5.8.3 Thermal-Aware Test Scheduling Results

Table 5.4 notes the results for thermal-aware PSO-based test scheduling policy noted in Section 5.7. Experimental results have been shown for the three SoC benchmarks for different number of I/O pairs, power budgets, and the values of the parameter W in expression 5.1. The value $W = 0$ is expected to produce results optimized toward test time minimization, while $W = 1$ is expected to perform thermal optimization, ignoring the test time values.

5.9 SUMMARY

In this chapter, test scheduling approaches have been discussed for network-on-chip (NoC) based system-on-chip (SoC) designs.

TABLE 5.4 Thermal-Aware Test-Scheduling Results

SoC	W	I/O Pairs					
		2		3		4	
		Test time	Peak Temperature (°C)	Test time	Peak Temperature (°C)	Test time	Peak Temperature (°C)
d695, Power budget 22W	0	9807	130.0	6933	161.7	5272	169.8
	0.2	12162	95.2	7953	111.4	8000	114.9
	0.5	15859	83.2	15771	87.0	12405	90.1
	0.8	31267	64.1	29538	65.1	24518	70.8
	1.0	65271	52.5	63722	55.5	59035	58.5
p22810, Power budget 26W	0	145098	251.0	98167	320.0	85453	417.0
	0.2	159998	180.3	123198	198.8	122787	200.7
	0.5	273404	125.9	182058	150.6	174498	152.2
	0.8	357894	100.2	356242	103.2	355968	108.4
	1.0	1034348	79.5	997029	84.5	787475	89.5
p93791, Power budget 27W	0	159188	253.0	110338	297.0	85967	351.0
	0.2	172728	197.7	146508	212.3	136898	229.8
	0.5	260850	146.6	235952	169.1	211576	157.3
	0.8	467926	107.5	400158	110.2	382531	114.43
	1.0	1167650	75.8	1121532	80.8	1116166	82.8

Both preemptive and non-preemptive core testing philosophies have been discussed with the flexibility to choose the frequency of operation for the cores. Test patterns and responses are transmitted through the underlying NoC infrastructure. Test schedule results have been presented for cases with varying numbers of input-output pairs. Trade-off has been presented between the test time and peak temperature values for NoCs.

REFERENCES

1. K. Manna, S. Singh, S. Chattopadhyay, I. Sengupta, "Preemptive Test Scheduling for Network-on-Chip using Particle Swarm Optimization", *International Symposium on VLSI Design and Test*, 2013, pp. 74–82.
2. K. Manna, P. Khaitan, S. Chattopadhyay, I. Sengupta, "Particle Swarm Optimization based Technique for Network-on-Chip Testing", *International Conference on Emerging Applications of Information Technology*, 2012, pp. 66–69.
3. K. Manna, C.S. Sagar, S. Chattopadhyay, I. Sengupta, "Thermal-Aware Preemptive Test Scheduling for Network-on-Chip based 3D ICs", *IEEE Computer Society Annual Symposium on VLSI (ISVLSI)*, 2016, pp. 529–534.
4. K. Manna, V.C. Reddy, S. Chattopadhyay, I. Sengupta, "Thermal-Aware Multifrequency Network-on-Chip Testing using Particle Swarm Optimization", *International Journal on High Performance Systems Architecture*, Vol. 5(3), 2015, pp. 141–151.

Index